Why Wind Turbines Are Just Swam Ass Fans

Ali Ballswhip

ISBN: 978-1-77961-535-0
Imprint: Telephasic Workshop
Copyright © 2024 Ali Ballswhip.
All Rights Reserved.

Contents

Introduction   1
Overview of Wind Turbines   1

Environmental Impact of Wind Turbines   15
Impact on Wildlife   15
Impact on Landscapes   28
Effects on Tourism and Recreation   31

Economic Considerations of Wind Turbines   37
Cost-Benefit Analysis of Wind Turbines   37
Impact on Property Values   50

Technical Limitations and Challenges of Wind Turbines   59
Wind Variability and Unpredictability   59

Bibliography   65
Noise and Health Concerns   67
Maintenance and Longevity   76

Case Studies and Controversies   87
Cape Wind Project   87
Offshore Wind Farms   94
Wind Energy vs Other Renewable Sources   102

Future Trends and Alternative Solutions   111
Technological Advances in Wind Turbines   111
Other Renewable Energy Sources   119
Energy Efficiency and Conservation   129

Index 139

Introduction

Overview of Wind Turbines

Definition of Wind Turbines

Wind turbines are large structures that harness the power of the wind to generate electricity. They consist of three main components: the tower, the blades, and the nacelle. The tower is a tall, sturdy structure that supports the entire wind turbine. It is typically made of steel and can reach heights of up to 300 feet or more.

The blades, usually three in number, are attached to a hub and are connected to the nacelle. The blades capture the kinetic energy in the wind and convert it into rotational motion. They are designed to be aerodynamic, allowing them to efficiently catch the wind and transfer its energy to the rotor.

The nacelle is the housing unit that contains the generator, gearbox, and other essential components of the wind turbine. It is usually located behind the blades and rotates with them. The nacelle ensures that the wind turbine continuously faces the wind, maximizing its energy capture.

Inside the nacelle, the wind's rotational energy is converted into electricity. The generator converts the mechanical rotational motion into electrical energy. This energy can then be used to power homes, businesses, and even entire communities, depending on the capacity and arrangement of the wind turbine.

Wind turbines operate on a simple principle: when the wind blows, the motion created by it rotates the blades. As the blades spin, the kinetic energy of the wind is transferred to the generator, which produces electricity. The electricity generated is then transported through power lines to where it is needed.

It is important to note that wind turbines are part of the broader field of renewable energy, aimed at reducing dependence on fossil fuels and mitigating climate change. By harnessing the power of the wind, wind turbines provide a clean and sustainable source of electricity.

Although wind turbines are commonly associated with large-scale wind farms, they can also be found on smaller scales, such as residential or community installations. These smaller wind turbines can help individuals and communities become more self-sufficient in their energy needs and contribute to a cleaner environment.

Wind Turbine Components

To better understand wind turbines, let's take a closer look at each of their main components.

1. **Tower:** The tower provides the structural support for the entire wind turbine. It needs to be strong and tall to ensure that the blades are placed high enough to capture as much wind as possible. The height of the tower is also important to minimize the impact of turbulence and maximize the wind speed, as wind tends to blow stronger at higher altitudes.

2. **Blades:** The blades are the most recognizable part of a wind turbine. They are designed to capture the kinetic energy in the wind and convert it into rotational motion. To achieve this, the blades are aerodynamically shaped and made of lightweight materials like fiberglass or carbon fiber. The length and shape of the blades determine the amount of wind energy that can be captured.

3. **Nacelle:** The nacelle houses the vital components that enable the conversion of wind energy into electricity. It contains the generator, gearbox, and controls. The generator converts the rotation of the blades into electrical energy, while the gearbox adjusts the rotational speed of the blades to optimize power generation. The controls monitor and regulate the turbine's performance, ensuring its safe and efficient operation.

4. **Foundation:** The foundation is responsible for anchoring the wind turbine to the ground. It needs to be strong and stable enough to withstand the dynamic loads and forces exerted by the wind. The type of foundation required depends on various factors, such as wind conditions, soil composition, and turbine size.

Types of Wind Turbines

There are two primary types of wind turbines: horizontal-axis wind turbines (HAWTs) and vertical-axis wind turbines (VAWTs).

Horizontal-Axis Wind Turbines (HAWTs): HAWTs are the most common type of wind turbine used today. They have a horizontal axis of rotation, meaning the blades rotate around a central hub parallel to the ground. With their blades facing into the wind, HAWTs are highly efficient at capturing wind energy. They can range in size from small residential turbines to large commercial installations.

Vertical-Axis Wind Turbines (VAWTs): VAWTs have a vertical axis of rotation, with the blades arranged in a helical or eggbeater-like configuration. Unlike HAWTs, VAWTs can capture wind from any direction, making them more suitable for urban or turbulent wind environments. However, VAWTs typically have lower efficiency compared to HAWTs and are less common in commercial applications.

How Wind Turbines Work

Wind turbines work on the principle of converting the kinetic energy in the wind into mechanical and electrical energy. Let's explore the step-by-step process:

1. **Wind Capture:** As the wind blows, the blades of the wind turbine start to rotate. The aerodynamic shape of the blades allows them to capture the wind's kinetic energy effectively. The wind flows over and under the blades, creating a difference in air pressure, which generates lift and causes the blades to spin.

2. **Rotor Rotation:** The rotational motion of the blades is transferred to the rotor, which is connected to the generator inside the nacelle. The rotor spins at a rotational speed proportional to the incoming wind speed.

3. **Generator Operation:** Inside the nacelle, the mechanical energy from the rotating rotor is converted into electrical energy by the generator. The generator contains a series of electromagnets and conductive coils that create a magnetic field. As the rotor spins, the magnetic field induces an electric current in the coils.

4. **Power Transmission:** The electricity generated by the wind turbine is transported through power lines to where it is needed. It can be used to power homes, buildings, or fed into the electrical grid for wider distribution.

5. **System Monitoring and Controls:** Wind turbines are equipped with monitoring and control systems to ensure their safe and efficient operation. These systems continuously monitor the turbine's performance, wind

conditions, rotor speed, and other critical parameters. They also facilitate communication with the grid and enable the turbine to respond to changes in wind speed or grid demand.

Key Factors Affecting Wind Turbine Performance

Several factors influence the performance and efficiency of wind turbines. Understanding these factors is essential for optimizing wind turbine design and operation. Let's explore the key factors:

- **Wind Speed and Direction:** Wind turbines require a minimum threshold wind speed to start turning and generate electricity. The wind speed also affects the power output of the turbine. Higher wind speeds result in increased power generation, up to a certain point where the turbine reaches its rated capacity. Wind direction is crucial too, as turbines need to align with the wind for optimal performance.

- **Blade Length and Design:** The length and design of the blades play a significant role in capturing wind energy. Longer blades have a larger swept area, allowing them to intercept more wind and generate more power. Blade design optimization involves considerations such as airfoil shape, twist, and angle of attack to maximize efficiency while minimizing noise and turbulence.

- **Capacity Factor:** The capacity factor quantifies the actual energy output of a wind turbine compared to its maximum potential output. It depends on various factors, including wind resource availability, downtime for maintenance and repairs, and limitations in gearbox or generator efficiency. Higher capacity factors indicate more efficient utilization of a wind turbine's rated capacity.

- **Environmental Conditions:** Environmental factors, such as temperature, humidity, and altitude, can affect wind turbine performance. Extreme temperatures and icing conditions can impact the functioning of different components, requiring additional design considerations and maintenance protocols in specific regions.

- **Wake Effects:** Wake effects occur when the wake of one wind turbine impacts the performance of nearby turbines. Turbines placed downwind may experience reduced wind speeds and turbulence due to the presence of an upstream wind turbine. Proper spacing and positioning of turbines can minimize wake effects and optimize overall wind farm performance.

- **Maintenance and Operations:** Regular maintenance and proper operation are essential to ensure the longevity and efficiency of wind turbines. These activities include ongoing monitoring, routine inspections, lubrication, cleaning, and timely repairs. Efforts are made to reduce maintenance costs and minimize downtime for optimal power generation.

- **Grid Integration:** Integrating wind turbines into the electrical grid poses challenges related to power stability, balancing supply and demand, and grid reliability. Advanced grid integration techniques, such as energy storage systems and smart grid technologies, help offset variability and intermittency challenges associated with wind power.

Real-World Applications of Wind Turbines

Wind turbines have become a prominent feature of the global energy landscape, with significant installations worldwide. Let's explore some real-world applications:

- **Onshore Wind Farms:** Onshore wind farms, consisting of multiple wind turbines erected on land, are commonly found in areas with favorable wind conditions. These wind farms range in size, from small community-based installations to large-scale projects comprising hundreds of turbines. They contribute to local and national energy grids, offsetting the use of conventional fossil fuels.

- **Offshore Wind Farms:** Offshore wind farms are located in bodies of water, typically in coastal areas or deep-sea environments. These wind farms harness the strong and consistent winds available at sea, making them ideal for large-scale power generation. Offshore wind farms are being developed globally and have the potential to provide significant amounts of clean energy.

- **Residential and Community-Scale Wind Turbines:** Smaller wind turbines can be found in residential areas and communities, providing an alternative power source for individual homes or shared spaces. These turbines help reduce dependence on the electrical grid, lower energy costs, and promote energy self-sufficiency.

- **Hybrid Energy Systems:** Wind turbines are often integrated into hybrid energy systems, combining wind power with other renewable energy sources like solar photovoltaics or energy storage. These systems provide a more stable and reliable energy supply, utilizing complementary resources and reducing the reliance on intermittent wind conditions.

- **Island and Remote Applications:** Wind power plays a significant role in powering islands and remote areas with limited access to conventional energy sources. Wind turbines can provide reliable and sustainable electricity, reducing the reliance on costly and environmentally damaging diesel generators or expensive energy imports.

- **Industrial and Commercial Installations:** Wind power is also used by various industries and commercial entities to meet their energy needs. Large-scale wind turbines can be found in industrial zones, manufacturing facilities, and commercial buildings, providing a renewable and cost-effective source of electricity.

Wind turbines offer a tangible and visible solution to address the pressing challenges of climate change and the transition to clean energy sources. They demonstrate the potential for the widespread adoption of renewable energy technologies and play a crucial role in achieving a sustainable future.

Historical Background of Wind Turbines

"Back in the day, before wind turbines were even a blip on the renewable energy radar, humans were already harnessing the power of the wind for various purposes. It all started with our ancestors, who probably noticed that the wind could make their lives a little bit easier. And let's face it, they needed all the help they could get, what with all the caveman drama and hunting and gathering."

"But let's fast forward a bit to Ancient Greece, where the true pioneers of wind power emerged. They were the first to use wind energy to sail and transport goods across the seas. Those Greeks really knew how to make the wind work for them, literally - they were the ones who invented the sailboat. Clever bunch, those Greeks. They knew the wind was a force to be reckoned with, and they used it to their advantage."

Windmills: A Marvel of Engineering

"Now, let's bring the spotlight to windmills. These magnificent structures became the wind power superheroes of the Middle Ages. Originating in the Middle East and Persia around the 9th century, windmills were a game-changer. They didn't just harness the

power of the wind for transportation, they put it to work grinding grains, sawing wood, and pumping water. Talk about multi-tasking!"

"Windmills quickly spread across Europe, becoming a fixture in the landscape. They were like the celebrity status symbols of their time. Everyone wanted a windmill! They were symbols of progress, innovation, and wealth. And let's not forget they were also a great photo opportunity - think windmill selfies before selfies were even a thing."

"Over time, windmill designs evolved and became more sophisticated. Engineers figured out that using multiple blades instead of just one increased efficiency. And let me tell you, these windmills were a sight to behold. They dotted the countryside, their giant arms gracefully turning in the wind, like synchronized dancers. It was truly a magical sight."

Modern Wind Turbines: From Humble Beginnings to Global Giants

"But let's fast forward to the 21st century, where wind turbines have taken the stage as the stars of renewable energy. These sleek, towering giants have made wind power not just a novelty, but a serious contender in the race to combat climate change. They are the eco-warriors we desperately need in our fight against fossil fuels."

"The concept of modern wind turbines, also known as wind turbines we see today, dates back to the late 19th century. Inventors such as Charles Brush and Thomas Edison played around with wind power, trying to harness it for electricity generation. But it wasn't until the mid-20th century that wind turbines really started to take off (pun intended)."

The Oil Crisis and the Birth of the Wind Energy Revolution

"In the 1970s, something big happened - the oil crisis hit. Suddenly, countries realized they needed to diversify their energy sources and reduce their dependence on fossil fuels. And thus, the wind energy revolution began. People started seeing wind turbines not just as a quaint relic of the past, but as a key player in the future of energy."

"Governments began investing in wind energy research and development, and engineers and scientists worked tirelessly to improve turbine designs and efficiency. It was an exciting time to be in the wind power industry - a time of innovation, exploration, and a whole lot of trial and error. You know what they say, you can't make an omelet without breaking a few wind turbine blades."

The Rise of Offshore Wind Farms

"As wind turbines grew in size and power, it became clear that land-based sites were not enough to meet the growing demand for renewable energy. That's when offshore wind farms entered the scene. These massive projects, situated in the vast open waters, promised even greater energy generation potential."

"Offshore wind farms have their unique set of challenges, from the engineering marvels required to withstand the harsh marine environment to the logistics of installation and maintenance. But they offer significant advantages too - the wind is stronger and more *consistent offshore, and the visual impact on the landscape is reduced. Plus, let's be honest, wind turbines rising from the sea just look downright cool."*

Conclusion

"So, there you have it - a whirlwind tour through the historical background of wind turbines. From the humble beginnings of harnessing the wind for sailing to the awe-inspiring windmills of the Middle Ages, and finally, the modern wind turbines that dot our landscapes and coastlines, wind power has come a long way."

"But the story doesn't end here. As technology continues to evolve, so will wind turbines. We'll see larger and more efficient turbines, innovations in floating and vertical axis designs, and even greater integration with other renewable energy sources. The future of wind power is bright, my friends. Pun definitely intended."

"So, buckle up, because the wind turbine ride is just getting started. Together, let's explore the impact, challenges, and potential solutions for these swam ass fans and continue to move towards a more sustainable and renewable future."

Importance of Wind Turbines in Renewable Energy

Renewable energy has become an increasingly important topic in recent years, as the need to transition away from fossil fuels and reduce greenhouse gas emissions becomes more pressing. Wind turbines play a vital role in renewable energy, providing a clean and sustainable source of power. In this section, we will explore the importance of wind turbines in the broader context of renewable energy.

The Need for Renewable Energy

Before we delve into the specific importance of wind turbines, let us first understand why renewable energy is crucial. Fossil fuels, such as coal, oil, and natural gas, have been the primary sources of energy for many decades. However, their extraction and burning release vast amounts of carbon dioxide into the atmosphere, contributing to climate change and environmental degradation.

The urgent need to mitigate the effects of climate change has led to a global consensus on the necessity of transitioning to renewable energy sources. By harnessing naturally replenishing resources like wind, solar, hydro, and geothermal power, we can reduce our dependence on fossil fuels and mitigate the negative impacts of climate change. Wind turbines have emerged as a key player in this transition, offering significant advantages in terms of scalability, cost-effectiveness, and environmental friendliness.

Advantages of Wind Turbines

Wind turbines offer several advantages that make them a critical component of renewable energy generation. Let's explore some of these advantages in detail:

1. **Abundant and Free Resource** Wind is an abundant resource available in many parts of the world. By harnessing the power of the wind, we can tap into a virtually limitless source of energy. Unlike finite fossil fuel reserves, wind energy is free and does not deplete with use. Wind turbines enable us to capture this energy and convert it into electricity, providing a sustainable and renewable power source.

2. **Clean and Environmentally Friendly** Wind energy is one of the cleanest forms of energy available. Unlike fossil fuels, wind turbines do not emit any greenhouse gases or air pollutants when generating electricity. This makes wind power a crucial tool in reducing carbon emissions and combating climate change. By choosing wind energy over fossil fuels, we can improve air quality, protect human health, and alleviate environmental impacts.

3. **Scalable and Flexible** One of the key advantages of wind turbines is their scalability. They can be installed as standalone units or grouped together in wind farms, ranging from a few turbines to hundreds or even thousands. This scalability allows wind energy systems to be tailored to specific energy needs, from powering individual homes to supplying electricity to entire cities. Additionally, wind turbines can be integrated into existing power grids, providing a flexible and reliable source of energy.

4. **Job Creation and Economic Stimulus** The wind energy sector has the potential to create a significant number of jobs and stimulate economic growth. As wind power continues to expand, there is a growing demand for professionals in areas such as engineering, manufacturing, construction, and maintenance of wind turbines. Moreover, the development of wind farms often leads to increased economic activity in rural communities, through investments, job opportunities, and local infrastructure development.

Challenges and Solutions

While wind turbines offer numerous advantages, they also face certain challenges. Understanding these challenges and finding suitable solutions is crucial for harnessing the full potential of wind energy. Let's discuss some of the key challenges and their corresponding solutions:

1. **Intermittency and Energy Storage** One of the main concerns with wind energy is its intermittent nature, as wind speeds fluctuate over time. This intermittency poses challenges for maintaining a consistent supply of electricity to the grid. However, advancements in energy storage technologies, such as batteries and pumped hydro storage, offer effective solutions. By storing excess energy during periods of high wind generation and releasing it during low wind periods, energy storage systems can help address the issue of intermittency.

2. **Grid Integration and Transmission** Wind farms are often located in remote areas, far from the end-users. This long-distance transmission of electricity introduces challenges in efficiently integrating wind energy into existing power grids. To address this, improvements in grid infrastructure, such as the development of smart grids and high-voltage direct current (HVDC) transmission lines, can facilitate the smooth integration and transmission of wind power.

3. **Public Perception and Acceptance** While wind turbines are generally regarded as a clean energy solution, they can face opposition from local communities due to aesthetic concerns or perceived negative impacts on wildlife. Engaging in transparent and inclusive public consultation processes, conducting thorough environmental impact assessments, and implementing appropriate mitigation measures can help address public concerns and ensure the successful deployment of wind turbines.

Conclusion

Wind turbines are an essential component of renewable energy systems, offering a clean, abundant, and scalable source of power. By harnessing the power of the wind, we can reduce our reliance on fossil fuels, mitigate climate change, and create a sustainable future. While wind turbines face certain challenges, ongoing technological advancements and strategic solutions provide a pathway to maximizing the benefits of wind energy. Embracing the importance of wind turbines in renewable energy is crucial for a greener and more sustainable world.

Note: The content provided in this section serves as an overview of the importance of wind turbines in renewable energy. For more comprehensive information and in-depth understanding, readers are encouraged to further explore the references and resources listed at the end of this book.

References:

1. American Wind Energy Association. (2021). *Wind Energy Basics.* Retrieved from `https://www.awea.org/wind-101/basics-of-wind-energy`.

2. Global Wind Energy Council. (2021). *Why Wind Matters.* Retrieved from `https://gwec.net/why-wind-matters/`.

3. International Renewable Energy Agency. (2019). *Renewable Power Generation Costs in 2018.* Retrieved from `https://www.irena.org/publications/2019/May/Renewable-power-generation-costs-in-2018`.

4. United States Department of Energy. (2021). *Wind Energy Technologies Office.* Retrieved from `https://www.energy.gov/eere/wind/wind-energy-technologies-office`.

Further Reading:

1. Archer, C. L., & Jacobson, M. Z. (2005). *Evaluating the impacts of large-scale wind power deployment on power system reliability: The case of the Bonneville Power Administration.* Renewable Energy, 30(13), 1933-1952.

2. Yang, H., Zhou, W., and Tian, H. (2015). *Renewable and Sustainable Energy Reviews,* 44, 466-481.

3. Zervos, A. (2019). *The Wind Energy Handbook: A Guide to Wind Power Engineering.* Wiley.

Having explored the importance of wind turbines in renewable energy, let's now move on to the environmental impact of these structures and their effects on wildlife, landscapes, and human health.

The Problem with Wind Turbines

Wind turbines, touted as a clean and renewable source of energy, have gained significant popularity in recent years. However, behind the veneer of sustainability, there are undeniable problems associated with these towering structures. While wind energy has its merits, it is crucial to acknowledge and address the challenges that wind turbines bring forth.

Intermittency and Unpredictability

One of the primary problems with wind turbines is their intermittent nature. Unlike traditional power plants that can generate electricity consistently, wind turbines heavily rely on wind speed to produce energy. Unfortunately, wind is not constant, and its variability poses a significant challenge to the reliability of wind power.

During periods of low wind, such as calm summer days or during lulls in weather systems, wind turbines produce little to no electricity. This intermittency can strain power grids and require backup power sources, such as fossil fuel-based plants, to compensate for the shortfall. The need for backup power not only diminishes the environmental benefits of wind energy but also introduces additional costs and complexities to the overall energy system.

Furthermore, wind speed can be highly unpredictable, leading to sudden fluctuations in power output. This unpredictability poses challenges for grid operators in balancing electricity supply and demand, potentially causing stability issues in the grid. Advanced forecasting techniques and grid management strategies are required to mitigate the impact of wind unpredictability on power systems.

Environmental Impact

While wind turbines are often hailed as environmentally-friendly, they are not without their ecological consequences. One of the most significant concerns is the impact on wildlife, particularly birds and bats. The rotating blades of wind turbines can pose a threat to flying animals, resulting in collisions and mortality. Additionally, the noise emanating from operating turbines can cause disturbance and disorientation to wildlife, further disrupting their habitats and natural behavior.

Marine life is not spared from the environmental impact of wind turbines either. Offshore installations can disturb marine ecosystems during construction and operation, affecting fish populations and their migratory patterns. The underwater sound generated by the turbines can also interfere with marine mammal communication and navigation.

Moreover, wind turbines have adverse effects on landscapes and scenic views. The sheer size and number of turbines required to generate significant amounts of electricity can create visual pollution, altering the natural aesthetics of rural areas and coastal regions. This visual impact can have repercussions for tourism, as wind farms may deter visitors seeking unspoiled and picturesque destinations.

Technical Limitations and Grid Integration

The technical limitations of wind turbines also present challenges for their widespread deployment. As mentioned earlier, wind variability and intermittency necessitate backup power sources and energy storage solutions to ensure a stable and reliable electricity supply. These additional requirements come with their own set of technical and economic complexities.

Noise pollution is another concern associated with wind turbines, particularly the low-frequency noise generated by their operation. While research is ongoing to investigate the potential health impacts of this noise, it remains a subject of debate and public concern. Striking a balance between wind energy generation and mitigating noise-related issues is a technical challenge that needs to be addressed.

Maintenance and longevity are critical factors in the performance of wind turbines. These structures are subjected to harsh environmental conditions and mechanical stresses, leading to wear and tear over time. Timely maintenance and component replacements are essential to ensure the turbines' efficiency and lifespan, but they can be challenging, especially in remote locations or offshore installations.

Furthermore, the decommissioning and waste management of wind turbines at the end of their operational life pose additional technical and environmental challenges. Proper disposal and recycling of turbine components are necessary to minimize the impact on landfills and promote a circular economy for renewable energy technologies.

Public Perception and Acceptance

Public opinion and acceptance play a crucial role in the success and proliferation of wind energy. While many individuals support renewable energy initiatives, the proximity of wind turbines to residential areas can elicit concerns and resistance.

Some people find the visual impact of wind farms intrusive and undesirable, leading to opposition to their construction. Moreover, concerns about potential health effects from noise or low-frequency vibrations attributed to wind turbines have emerged in public discourse. Though scientific studies have not definitively established a direct causal link, addressing these perceptual challenges and building public trust are essential for the long-term acceptance and adoption of wind energy.

Conclusion

Wind turbines have the potential to contribute significantly to the global shift towards sustainable and clean energy sources. However, it is essential to recognize and address the problems associated with wind turbines to achieve a balanced and effective approach to renewable energy.

Finding solutions to mitigate the intermittent nature of wind power, minimize the environmental impact on wildlife and landscapes, overcome technical limitations and grid integration challenges, and addressing public perception and acceptance are fundamental steps in maximizing the benefits of wind energy while minimizing its drawbacks. With further advancements in technology, strategic planning, and public engagement, wind turbines can become a more efficient and widely accepted part of our renewable energy future.

Environmental Impact of Wind Turbines

Impact on Wildlife

Threats to Birds and Bats

As wind turbines continue to rise on the horizon, there is growing concern about the impact they have on birds and bats. These majestic creatures, part of our natural ecosystem, are facing serious threats due to the proliferation of wind farms. In this section, we will explore the significant risks wind turbines pose to avian and bat populations, touching on the factors contributing to these threats and potential solutions to mitigate their impact.

The Impact on Birds

Birds, with their ability to fly freely across vast distances, are particularly susceptible to the dangers presented by wind turbines. The spinning blades of these massive structures become a deadly obstacle for our feathered friends, leading to high rates of collision fatalities. Let's examine some of the key reasons why birds are at risk:

Migration Patterns and Collision Risks Millions of birds embark on epic migratory journeys each year, flying thousands of kilometers to reach their breeding grounds or wintering habitats. Unfortunately, many of these flight paths intersect with wind turbine installations. Birds are often attracted to the open areas surrounding turbines, mistaking them for potential perches or nesting sites. As a result, they experience an increased chance of colliding with the rotating blades, leading to severe injury or death.

Species Vulnerability Certain bird species are more vulnerable to wind turbine strikes than others. Birds with larger wingspans, such as eagles, hawks, and vultures, are at a greater risk due to their size making them more likely to come into contact with the blades. Additionally, migratory birds that fly at night, such as songbirds and waterfowl, are particularly prone to collisions as they navigate through unfamiliar territories in low light conditions.

Displacement from Habitat The installation of wind turbines can result in the displacement of bird populations from their natural habitat. Construction activities, noise, and disturbance associated with wind farms lead to altered nesting patterns and disrupted feeding grounds. These disruptions can have lasting ecological consequences, affecting population dynamics and overall biodiversity in the surrounding areas.

Effects on Marine Birds Coastal wind farms, erected on or near shorelines, pose additional threats to marine bird species. Seabirds, such as gannets, cormorants, and pelicans, rely on these areas for foraging and breeding. The presence of wind turbines alters their foraging patterns and can cause collisions with the rotating blades. Such disruptions can have a cascading effect on the entire marine ecosystem, impacting fish populations and other marine species that depend on these birds for food.

Mitigation Strategies Addressing the threats to birds posed by wind turbines requires a multi-pronged approach that combines careful planning, technology advancement, and environmental stewardship. Here are some potential strategies to mitigate these risks:

- **Strategic Placement of Wind Farms:** By taking into account bird migration patterns and avoiding key flyways or known breeding grounds, wind farms can be situated in locations that minimize the risk of collisions.

- **Avian Radar and Monitoring Systems:** Employing advanced radar technology can detect bird movements in real-time, allowing turbine operators to temporarily shut down turbines during peak migration periods or when high concentrations of birds are detected in the vicinity.

- **Reduction of Blade Reflectivity:** Coating the turbine blades with specialized materials that minimize reflectivity can reduce the likelihood of birds mistaking them for objects in their flight path.

- **Noise Dampening Techniques:** Developing quieter wind turbine technologies can help minimize noise pollution that can disrupt bird communication and nesting behaviors.

- **Post-construction Monitoring and Research:** Continuous monitoring of bird populations in the vicinity of wind farms contributes to our understanding of their long-term impact. This data can inform efforts to refine mitigation strategies and improve future wind turbine siting.

- **Education and Awareness Campaigns:** Raising public awareness about the threats wind turbines pose to birds can foster greater conservation efforts and support for the development of more bird-friendly turbine designs.

While these mitigation strategies can help address the threats to birds, it is crucial to recognize that continued research, collaboration between stakeholders, and ongoing monitoring are necessary to refine these approaches and ensure the preservation of avian species in the face of expanding wind energy production.

The Impact on Bats

Bats, often misunderstood and underappreciated creatures, also face significant risks due to wind turbines. These nocturnal flyers play a vital role in our ecosystems, as they consume vast quantities of insects, including agricultural pests. However, several factors make bats particularly vulnerable to the presence of wind farms:

Collisions and Barotrauma Similar to birds, bats can collide with the rotating blades of wind turbines, resulting in mortal injuries. Due to their sonar-based navigation, bats may have difficulty detecting the rapidly spinning blades or may be unable to react quickly enough to avoid them. In addition to direct collisions, the sudden changes in air pressure around the blades can cause barotrauma, internal injuries resulting from the difference in pressure between their lungs and the low-pressure areas created by the blades.

Habitat Disruption The construction and operation of wind farms can lead to habitat fragmentation and disturbance for bats. The clearing of land and installation of turbines can disrupt roosting sites, causing bats to abandon or avoid these areas. The noise generated by wind turbines may also interfere with bat echolocation, affecting their foraging abilities and potentially leading to reduced prey availability.

Barriers to Migration Bats, like birds, undertake seasonal migrations, often covering long distances to reach their preferred habitats. Wind farms situated along these migration routes can act as physical barriers, forcing bats to navigate perilous paths between the spinning blades. This can result in increased collision risks and energy expenditure during these critical periods.

Mitigation Strategies To mitigate the impact on bats, a combination of precautionary measures and adaptive management strategies can be employed. Here are some potential approaches to consider:

- **Imposing Seasonal Restrictions:** Implementing seasonal shutdowns during periods of high bat activity, such as the bat mating season or peak migration periods, can help minimize collisions.

- **Ultrasonic Deterrents:** Utilizing ultrasonic acoustic deterrents near wind turbines can create a sonic barrier that deters bats from entering the danger zone.

- **Turbine Blade Mitigation:** Developing turbine designs that reduce blade rotation during low-wind periods, when bats are most active, can reduce the likelihood of collisions.

- **Artificial Roosting Structures:** Creating alternative roosting sites near wind farms can compensate for the loss of natural roosting areas, reducing the disruption to bat populations.

- **Bat-friendly Lighting:** Employing lighting systems on wind turbines that are less attractive to insects can reduce bat activity in these areas, minimizing the risk of collisions as they pursue their prey.

- **Research and Monitoring:** Continued research on bat behavior and movements in relation to wind farms is crucial for improving our understanding of their vulnerabilities. This knowledge can inform adaptive management strategies that are tailored to specific bat species, ensuring their continued conservation.

By implementing these mitigation strategies, we can strike a balance between the need for clean energy and the protection of bats, ensuring their survival while harnessing wind power sustainably.

As we delve into the complex web of environmental and economic considerations surrounding wind turbines, it is essential to acknowledge the challenges and potential consequences they present to birds and bats. By understanding these threats and working collectively to address them, we can shape a future where renewable energy and conservation go hand in hand.

Disruption of Wildlife Habitat

Wind turbines have often been hailed as a solution to our energy needs, but the truth is they come with their own set of problems. One of the most significant issues is the disruption of wildlife habitat. While it's true that wind turbines don't emit greenhouse gases like fossil fuel power plants, their placement and operation can have negative impacts on the surrounding environment and wildlife.

When wind turbines are installed, they require vast areas of land to be cleared, resulting in the destruction of natural habitats. This clearing of land not only removes trees and plants that provide food and shelter for various species but also disrupts the delicate balance of the ecosystem. Animals that rely on these habitats for survival are forced to relocate or find alternative resources, putting additional stress on already struggling populations.

The construction and operation of wind turbines also pose risks to wildlife, particularly to birds and bats. The spinning blades of wind turbines present a significant hazard to flying animals, causing collisions that can be fatal. Birds, in particular, are prone to flying into the path of the rotating blades, leading to injury or death. Bats, on the other hand, are susceptible to barotrauma, a condition caused by the rapid changes in air pressure near the moving blades, which can damage their respiratory systems and cause internal bleeding.

Efforts have been made to mitigate these risks by studying bird migration patterns and implementing measures to reduce bird fatalities. These measures include proper siting of wind turbines away from important migratory routes, adjusting turbine operations during peak migration periods, and using avian detection systems to detect and deter birds from entering dangerous areas. However, these solutions are not foolproof and don't eliminate the problem entirely.

The disruption of wildlife habitat caused by wind turbines extends beyond the direct impacts on birds and bats. The noise generated by the turbines can disturb and displace other wildlife species, such as nesting birds and mammals. The constant hum and vibration can interfere with their communication, breeding, and foraging behaviors. Additionally, the infrastructure associated with wind farms, including access roads and transmission lines, can fragment habitats, making it difficult for animals to move freely and find food and mates.

Marine life is not immune to the effects of wind turbines either. Offshore wind farms can alter the underwater soundscape, affecting marine mammals and fish that rely on sound for navigation, communication, and finding prey. The underwater noise generated by the construction and operation of wind turbines can disorient marine species and disrupt their natural behavior patterns.

The impact of wind turbines on endangered species adds another layer of complexity to the issue. Endangered species are often particularly vulnerable to habitat disruption, and the placement of wind farms in their habitats can further threaten their survival. The potential harm caused to these already vulnerable populations must be carefully considered and weighed against the benefits of renewable energy.

While wind turbines offer a cleaner alternative to fossil fuels, their negative impact on wildlife habitat cannot be ignored. Striking a balance between renewable energy generation and the preservation of wildlife habitats is a complex challenge that requires careful planning, thorough impact assessments, and ongoing research.

It's crucial for policymakers, renewable energy developers, and environmental organizations to collaborate and find innovative solutions that minimize the disruptions to wildlife habitat while still supporting the transition to renewable energy sources.

To explore the complexities of this topic further, consider the following questions:

1. How can the siting of wind turbines be optimized to minimize the disruption of wildlife habitats? 2. What additional measures can be implemented to further mitigate the risk of bird and bat collisions with wind turbine blades? 3. How can the noise pollution caused by wind turbines be reduced to minimize disturbances to wildlife? 4. Are there alternative renewable energy sources that have a lesser impact on wildlife habitat? 5. How can we incentivize the development of renewable energy technologies that take into account the conservation of wildlife habitats?

As we continue to seek greener energy options, it's essential to remember that the preservation of our natural environment and its inhabitants should remain a top priority.

Noise Pollution and Disturbance

Noise pollution is a significant concern associated with the operation of wind turbines, which can cause disturbances and have negative impacts on both humans and wildlife. In this section, we will explore the sources of noise from wind turbines, its effects on different populations, and potential solutions to mitigate this problem.

Sources of Noise

Wind turbines generate noise through various mechanisms, including aerodynamic interaction, mechanical components, and electrical processes. The main sources of noise can be categorized as follows:

1. **Aerodynamic Noise:** As wind flows over and around the blades, it causes turbulence and aerodynamic pressure fluctuations, leading to noise generation. The interaction between the blades and the air produces both tonal and broadband noise components.

2. **Mechanical Noise:** Mechanical components within the turbine, such as the gearbox and generator, can produce noise due to the rotation and interaction of various machinery parts. These noises are often tonal and low-frequency in nature.

3. **Electrical Noise:** Electrical components, such as inverters and transformers, can introduce high-frequency harmonic noise into the system.

The noise produced by wind turbines can be further influenced by factors such as wind speed, turbine size, and operational conditions. It is important to assess and manage these noise sources to minimize potential disturbances.

Effects on Humans

Excessive noise from wind turbines can have adverse effects on human health and well-being. Prolonged exposure to high levels of noise can lead to various physical and psychological impacts, including:

- **Sleep Disturbance:** Wind turbine noise, especially the low-frequency components, can disrupt sleep patterns, leading to sleep deprivation and fatigue. This can have cascading effects on overall health and cognitive functioning.

- **Annoyance and Stress:** The continuous noise emitted by wind turbines can result in annoyance and stress among individuals living in close proximity to wind farms. This can negatively affect quality of life and mental well-being.

- **Health Issues:** Some studies suggest that prolonged exposure to wind turbine noise may contribute to the development of cardiovascular diseases, such as hypertension and heart problems. However, more research is needed to establish a conclusive link.

To address these concerns, regulatory bodies often impose noise limits to ensure that wind turbines operate within acceptable noise thresholds. It is crucial to strike a balance between renewable energy generation and the well-being of the local communities.

Effects on Wildlife

Wind turbines can also have detrimental effects on wildlife, specifically on bird and bat populations. The rotating blades of wind turbines can pose a collision risk and cause direct mortality. Additionally, the noise generated by wind turbines can have indirect effects on wildlife, including:

- **Disruption of Communication and Foraging:** High noise levels can interfere with animal communication, making it harder for them to detect and respond to important signals. This can impact mating rituals, hunting, and foraging activities, leading to decreased reproductive success and food availability.

- **Habitat Displacement:** Wildlife may alter their behaviors and movements to avoid areas with high noise levels, leading to displacement from their natural habitats. This can result in habitat fragmentation and ecosystem imbalances.

- **Stress and Physiological Effects:** Prolonged exposure to noise can induce stress responses in wildlife, affecting their immune systems, reproductive success, and overall health. These effects can have long-term implications for population dynamics.

Conservation efforts and proper siting of wind turbines are crucial in minimizing the impacts on wildlife. Studying the migration patterns and behavioral habits of sensitive species can guide the placement of wind farms to mitigate these effects.

Mitigation Strategies

To address the issue of noise pollution and disturbance caused by wind turbines, several mitigation strategies can be employed:

- **Noise Barrier Design:** Installing physical barriers, such as sound-attenuating walls, around wind turbines can reduce the propagation of noise, especially in the direction of sensitive receptors. They act as a shield, absorbing or reflecting sound waves and reducing the overall impact on nearby communities.

- **Turbine Design and Placement:** Improved turbine design and placement can help minimize noise generation. Factors such as blade shape, aerodynamic profiles, and tower height can be optimized to reduce aerodynamic and mechanical noise. Furthermore, locating wind turbines away from noise-sensitive areas can help mitigate disturbances.

- **Noise Monitoring and Compliance:** Regular monitoring of noise emissions from wind turbines is essential to ensure compliance with established regulations and standards. Compliance testing can help identify potential sources of noise and guide the implementation of mitigation measures if necessary.

- **Community Engagement:** Engaging with local communities during the planning and development stages of wind farms can help address concerns related to noise pollution. Open communication, education, and involvement in decision-making processes can foster a sense of ownership and reduce potential conflicts.

It is important to adopt a holistic approach that balances the benefits of renewable energy with the need to minimize noise pollution. Continued research, technological advancements, and community involvement are essential in mitigating the impact of wind turbine noise on both humans and wildlife.

Summary

Noise pollution and disturbance associated with wind turbines have significant implications for both human populations and wildlife. The noise generated by wind turbines can lead to sleep disturbances, annoyance, stress, and potential health issues for nearby residents. Additionally, wildlife can be negatively impacted through direct mortality as well as disruption of communication, foraging, and habitat displacement. Mitigation strategies such as noise barrier design, turbine design optimization, compliance monitoring, and community engagement can help address and minimize these impacts. Striking a balance between renewable energy generation and the well-being of local communities and ecosystems is crucial for a sustainable future.

Effects on Marine Life

Wind turbines, those swam ass fans, may be touted as a clean and renewable source of energy, but they also come with a hefty price for our marine friends. These towering structures have significant impacts on marine life, ranging from direct physical obstacles to disruptions in their natural habitat. In this section, we will explore the various effects that wind turbines have on our underwater neighbors.

Collision Hazards for Marine Animals

One of the most evident and tragic consequences of wind turbines for marine life is the risk of collisions. The rotating blades of wind turbines can pose a severe threat to birds, bats, and, yes, even our marine buddies. Flying creatures, such as seabirds and migratory birds, might mistake the spinning blades for an open passage and end up colliding with them. Similarly, diving seabirds that submerge while hunting for fish can get caught in the underwater foundations of offshore wind turbines, leading to injuries or fatalities.

But it's not just birds that fall victim to the merciless blades. Marine mammals, such as seals, whales, and dolphins, are also at risk of collision, particularly in areas where they are known to migrate or gather. These collisions can cause severe injuries or even death, as the massive blades can crush or maim these unsuspecting creatures.

Disruption of Underwater Ecosystems

Wind turbines may also disrupt the fragile balance of underwater ecosystems, causing cascading effects on marine life. The construction and operation of offshore wind farms involve extensive underwater activities, such as pile driving, dredging, and cable laying. These activities introduce high levels of noise and vibrations, which can disorient and stress marine organisms.

The noise pollution generated by wind turbines has adverse effects on marine mammals that rely on underwater sound for communication, navigation, and locating prey. Whales, for example, possess highly sensitive hearing capabilities, and the constant noise from wind turbines can interfere with their natural behaviors, including feeding and mating.

In addition to noise pollution, the electromagnetic fields generated by the cables connecting offshore wind turbines to the power grid can also disrupt the sensory perception and navigation abilities of certain marine species. For instance, some fish and sea turtles rely on Earth's magnetic fields for orientation during migration.

Alteration of Marine Habitat

The installation of wind turbines in marine environments requires large-scale modifications to the seabed and surrounding areas. During construction, the seabed is often dredged to create stable foundations for the turbines. These dredging activities can destroy or disrupt critical habitats, such as seagrass beds, coral reefs, and other underwater structures that serve as nurseries for countless marine species.

Furthermore, the deployment of wind turbines in coastal areas can introduce artificial structures that attract organisms, altering the natural community composition and biodiversity. For example, the turbine bases can serve as artificial reefs, attracting fouling organisms and potentially altering the local food web dynamics.

Effects on Fish and Marine Mammal Migration

Wind turbines, particularly those deployed offshore, can obstruct the migratory routes of fish and marine mammals. Many species rely on these migration pathways for breeding, feeding, and survival. The presence of wind turbines can impede these movements and disrupt critical life stages.

In some cases, wind farms can act as barriers, blocking the access of fish to their spawning grounds or preventing their migration to feeding areas. This disruption to vital processes can have significant consequences for population dynamics and the overall health of marine ecosystems.

Challenges of Mitigation and Conservation

Mitigating the adverse effects of wind turbines on marine life presents significant challenges. Developing strategies to reduce the number of collisions between wildlife and turbine blades is a priority. This may involve conducting thorough environmental impact assessments before constructing wind farms, identifying high-risk areas for wildlife, and implementing measures such as proper siting of turbines and improved lighting to minimize collision risks.

Moreover, employing advanced monitoring technologies can help assess the impacts of wind turbines on marine life and guide conservation efforts. For instance, acoustic monitoring can provide insights into the underwater soundscape and the responses of marine mammals to wind turbine noise.

Conservation organizations and researchers also play a crucial role in studying the long-term effects of wind turbines on marine ecosystems, including their cumulative impacts and potential interactions with other human activities, such as shipping and fishing.

Conclusion

As we strive to transition to cleaner and more sustainable sources of energy, it is essential to recognize and address the potential negative consequences that wind turbines can have on marine life. By understanding the effects on our underwater neighbors and working towards mitigating these impacts, we can strike a balance between renewable energy generation and the preservation of marine ecosystems.

Remember, when it comes to wind turbines and marine life, we need to find a solution that doesn't make the marine creatures wind up as casualties in the battle against climate change. So, let's push for innovation and better practices in wind farm development to ensure a harmonious coexistence between green energy and our marine friends.

Impact on Endangered Species

Wind turbines, touted as environmentally friendly sources of renewable energy, have raised concerns about their potential impact on endangered species. While they may help reduce greenhouse gas emissions and combat climate change, their presence and operation can pose various threats to vulnerable wildlife populations. In this section, we will explore the potential impacts of wind turbines on endangered species and discuss ways to mitigate these risks.

Threats to Avian Species

One of the primary concerns regarding wind turbines is their potential to harm avian species, including endangered birds. Collisions with turbine blades can result in fatalities, especially for birds that fly at night or migrate over long distances. Certain species, such as raptors, are particularly at risk due to their large size and hunting behaviors. Additionally, the spinning blades can create a visual flicker effect that disorients birds, increasing the likelihood of collisions.

To address this issue, various strategies can be implemented. Advanced radar and lidar technologies can be used to detect bird movements and trigger temporary shutdowns of turbines during peak migration periods. Furthermore, careful siting of wind farms away from known bird migration routes and sensitive habitat areas can minimize the risk of collisions. Environmental impact assessments prior to turbine installation can identify potential risks and guide mitigation efforts.

Impacts on Bat Populations

While birds are often the focus of attention, bats are also significantly impacted by wind turbines. These nocturnal insect-eating creatures are known to be particularly susceptible to fatalities due to their tendency to fly in close proximity to turbines. The low-pressure region created by spinning blades can cause barotrauma, leading to internal injuries or death.

Mitigation measures aimed at protecting bats often involve adjusting turbine operations during peak activity periods. Implementing curtailment strategies, such as reducing turbine speed or blade pitch during specific times, can significantly reduce bat fatalities. Additionally, studies suggest that incorporating ultrasonic deterrents into turbine design or locating wind farms away from important bat roosting areas can mitigate the risks to these endangered species.

Loss and Fragmentation of Habitat

Apart from direct impacts on individual species, wind turbines can also result in the loss and fragmentation of habitat, which can negatively affect endangered species. The construction of wind farms necessitates clearing land and installing access roads and transmission lines, resulting in the destruction and disruption of ecosystems.

To mitigate this issue, developers can employ land-use planning strategies that prioritize the use of existing disturbed lands or low-quality agricultural lands for wind turbine installation. Conservation easements or habitat restoration programs can also be implemented to compensate for the habitat loss incurred during construction. By carefully selecting sites and employing mitigation methods, it is possible to minimize the impact on endangered species and their habitats.

Noise Pollution and Disturbance

In addition to physical impacts, wind turbines generate noise that can disturb endangered species. Species sensitive to noise, such as marine mammals like whales and dolphins, may be affected by the underwater noise produced by offshore wind turbines. This disturbance can disrupt their communication, feeding, and breeding behaviors, potentially impacting their survival and reproductive success.

To mitigate the impact of noise pollution, implementing setback distances between wind farms and sensitive habitats can be crucial. The use of underwater noise-reducing technologies, such as bubble curtains or quieter turbine designs, can also help minimize the disruption to marine life. Environmental monitoring programs can be established to track the noise levels and potential impacts on endangered species, enabling adaptive management strategies to be implemented if necessary.

Long-Term Monitoring and Adaptive Management

To ensure the continued protection of endangered species, long-term monitoring and adaptive management strategies are essential. Regular monitoring programs can assess the effectiveness of mitigation measures and identify any unforeseen impacts on endangered species. This allows for adjustments and improvements to be made, ensuring that wind turbines operate in a manner that minimizes harm to these vulnerable populations.

Furthermore, collaboration between experts in ecology, conservation, and renewable energy can help facilitate the development of best practices and guidelines for minimizing the impact of wind turbines on endangered species. By applying scientific knowledge, technological advancements, and innovative approaches, it is possible to strike a balance between renewable energy generation and species conservation.

Conclusion

While wind turbines play a crucial role in the transition towards a cleaner and more sustainable energy future, their potential impact on endangered species cannot be overlooked. By recognizing the potential threats and implementing appropriate mitigation measures, we can minimize harm to vulnerable wildlife populations. Through continued research, innovation, and collaboration, we can strive for a future where renewable energy and biodiversity conservation go hand in hand. Let us not forget the importance of protecting endangered species, even in our pursuit of a greener world.

Impact on Landscapes

Visual Pollution

When discussing the impact of wind turbines, one cannot overlook the issue of visual pollution. While proponents of wind energy applaud the sight of towering turbines dotting the landscape, there are many who find them to be eyesores that disrupt the natural beauty of their surroundings. In this section, we will delve into the various aspects of visual pollution caused by wind turbines and explore the implications it has on individuals, communities, and tourism.

Visual Impact of Wind Turbines

One of the first things that comes to mind when thinking about wind turbines is their towering height and prominent presence. These giant structures can reach up to 100 meters or more, and their rotating blades can span anywhere from 50 to 80 meters. As a result, wind turbines can have a significant visual impact on the landscape, especially when clustered together in wind farms.

The sheer size and number of wind turbines can dominate the skyline and drastically alter the aesthetic appeal of an area. For example, imagine standing on a pristine hillside, enjoying the panoramic view of rolling hills and open skies, only to have that view obstructed by a row of imposing wind turbines. The contrast between nature's beauty and man-made structures can be jarring and disheartening for those who value their surroundings.

Effects on Tourism and Recreation

Visual pollution caused by wind turbines can have far-reaching consequences for tourism and recreational activities. Many areas rely heavily on their natural beauty and scenic landscapes to attract tourists. Wind turbines, with their industrial appearance, can deter visitors who seek a more traditional or unspoiled experience.

Tourism-dependent communities, such as coastal towns or rural areas known for their picturesque landscapes, may see a decline in visitor numbers due to the presence of wind turbines. This, in turn, can have a negative impact on the local economy, as businesses that cater to tourists may suffer from decreased revenue.

Furthermore, recreational activities such as hiking, camping, or birdwatching may be affected by the visual pollution caused by wind turbines. Outdoor enthusiasts who seek refuge in nature may find it harder to escape the encroachment of human activity with these large structures dominating the horizon.

Cultural and Historical Preservation

Besides the tangible impact on tourism and recreation, wind turbines can also pose challenges in terms of cultural and historical preservation. Many regions are known for their cultural heritage or historical significance, and the introduction of wind turbines can disrupt the integrity of these sites.

For example, imagine a quaint village that has stood for centuries, its charm derived from its intact historical buildings and picturesque streetscapes. The installation of wind turbines nearby can detract from the village's character and heritage, diminishing its appeal to visitors and potentially erasing a part of its identity.

Preserving cultural and historical landmarks is vital for maintaining a sense of identity and heritage. Wind turbines, when not thoughtfully placed, can undermine these efforts and interfere with the efforts to ensure the longevity of these valuable sites.

Addressing the Issue

While the visual pollution caused by wind turbines is undeniable, it is essential to consider both sides of the argument. Supporters of wind energy argue that the aesthetic impact of renewable energy sources must be weighed against the potential environmental and economic benefits they offer.

However, striking a balance between clean energy and preserving the visual integrity of landscapes is possible through strategic planning and decision-making. Governments and stakeholders can collaborate to implement measures that aim to minimize the visual impact of wind turbines.

One approach is to consider the placement of wind farms in locations where they can blend with the surroundings, such as areas with existing industrial infrastructure or near highways. Additionally, implementing zoning regulations and guidelines can help ensure that wind turbines are strategically placed, minimizing their impact on scenic areas.

Furthermore, advancements in technology offer opportunities to develop innovative design solutions. For instance, researchers are exploring the possibility of integrating wind turbines into existing structures or creating designs that are aesthetically appealing, such as turbines that resemble traditional windmills.

By addressing the issue of visual pollution proactively and involving all stakeholders in the decision-making process, it is possible to strike a balance between the critical need for renewable energy and the preservation of natural beauty.

Conclusion

Visual pollution caused by wind turbines is a contentious issue that cannot be ignored. The towering height and industrial appearance of wind turbines can disrupt the natural beauty of landscapes, impacting individuals, communities, and tourist destinations.

The imposition of wind turbines on scenic areas can have adverse effects on tourism, recreational activities, and cultural preservation. However, through strategic planning, innovative design, and stakeholder involvement, it is possible to mitigate the visual impact of wind turbines while still harnessing the power of renewable energy.

Finding a balance between clean energy and preserving the visual appeal of landscapes is vital to ensure a sustainable future that values both environmental protection and the preservation of our natural and cultural heritage.

Effects on Tourism and Recreation

Introduction

Wind turbines, with their towering structures and rotating blades, can have a significant impact on the tourism and recreation industries. While wind energy is lauded as a clean and renewable source of power, the visual presence and noise generated by wind turbines can disrupt the tranquility of natural landscapes and reduce the appeal of tourist destinations. In this section, we will explore the various effects of wind turbines on tourism and recreation, including visual pollution, impacts on outdoor activities, and potential solutions to mitigate these concerns.

Visual Pollution: The Blemish on Natural Landscapes

One of the major concerns regarding wind turbines is their visual impact on scenic landscapes. The sheer size and number of turbines in wind farms can disrupt the aesthetic beauty of natural environments, altering the character of tourist destinations. Imagine standing on the shores of a serene lake, only to be greeted by a sprawling array of towering wind turbines on the horizon. This visual intrusiveness can negatively affect the overall experience for visitors, diminishing the appeal of the area.

To illustrate this point, let's consider a real-world example: the Cape Cod region in Massachusetts, USA. This picturesque coastal area has long been a popular tourist destination known for its pristine beaches and charming seaside towns. However, the proposed Cape Wind Project, which planned to install 130 wind turbines off the coast, sparked intense opposition from residents and environmentalists alike. They argued that the turbines would mar the scenic beauty and compromise the unique character of the region, potentially driving tourists away.

Impacts on Outdoor Activities

The presence of wind turbines can also interfere with various outdoor activities, reducing the attractiveness of tourism and recreational areas. For instance, some tourists seek solace in birdwatching, enjoying the serene sights and sounds of avian species in their natural habitats. Unfortunately, wind turbines pose a threat to bird populations due to collision risks and habitat disruption. The spinning blades can be fatal for birds, especially those that migrate or inhabit areas near wind farms. This has raised concerns among bird enthusiasts, who worry about the potential decline in bird diversity within affected regions.

Additionally, outdoor recreational activities such as hiking, camping, and fishing may be negatively affected by wind turbine installations. The noise generated by the rotating blades can disturb the tranquility of natural environments, detracting from the overall experience. Some tourists may prefer peaceful retreats away from the modern world, seeking solace in the stillness and quietude of nature. The constant hum of wind turbines can shatter this desired serenity, leading to a decline in visitor satisfaction.

Mitigation Measures: Finding a Balance

While the impacts of wind turbines on tourism and recreation are significant, there are several measures that can be taken to mitigate these concerns and strike a balance between renewable energy generation and maintaining the attractiveness of tourist destinations.

One approach is careful site selection and planning. Wind farms should be strategically located away from highly frequented tourist spots, nature reserves, and areas of ecological importance. By minimizing the visual impact on scenic landscapes and the disturbance to wildlife habitats, the negative effects on tourism and recreation can be reduced.

Moreover, innovative design and technology can play a role in reducing the visual and acoustic impacts of wind turbines. For example, advancements in blade design can result in quieter operation, minimizing noise pollution in the surrounding areas. Furthermore, the integration of wind turbines with other structures, such as buildings or infrastructure, can help camouflage their presence, blending them into the surrounding environment more effectively.

Conclusion

Wind turbines have the potential to provide significant amounts of clean and renewable energy, but their presence can also create challenges for the tourism and recreation sectors. The visual intrusion on natural landscapes, disruption of outdoor activities, and concerns about wildlife preservation are valid factors that cannot be ignored. However, through thoughtful planning, site selection, and technological advancements, we can mitigate these impacts and find a harmonious balance between renewable energy generation and the preservation of our cherished tourist destinations. It is crucial to continue exploring innovative solutions to ensure a sustainable future that prioritizes both clean energy and the preservation of our natural heritage.

Cultural and Historical Preservation

Cultural and historical preservation is an essential consideration when discussing the impact of wind turbines on landscapes. Wind turbines have the potential to disrupt and even destroy cultural and historical sites that hold significant value for communities and humanity as a whole.

Preservation of Cultural Heritage

Cultural heritage refers to the traditions, customs, artifacts, and buildings that are inherited from past generations. It encompasses the unique identity and history of a community or society. Wind turbines, when situated in or near culturally significant areas, can pose a threat to the preservation of this heritage.

One of the main concerns is the visual impact of wind turbines on historical landscapes and buildings. Many historical sites rely on their scenic views and tranquility to evoke a sense of the past. The introduction of large wind turbines can disrupt the aesthetic harmony and diminish the authenticity of these sites. It is crucial to strike a balance between renewable energy goals and preserving the cultural value of such locations.

Challenges and Mitigation

Preserving cultural heritage in the context of wind turbines requires careful planning and consideration. Here are some challenges and potential mitigation strategies:

1. **Site Selection:** Rigorous assessments and consultations with cultural heritage experts should be carried out before the installation of wind turbines. Certain landscapes and areas hold unique historical, archaeological, or traditional significance and should be avoided to prevent any adverse impacts. By involving relevant stakeholders early in the planning process, potential conflicts between renewable energy and cultural preservation can be identified and addressed.

2. **Visual Impact Assessments:** Before implementing wind energy projects near culturally or historically valuable sites, a comprehensive visual impact assessment should be conducted. This assessment should consider viewpoints, sightlines, and the overall integrity of the landscape. It is essential to find a balance between harnessing clean energy and respecting the cultural, aesthetic, and historical values of the surroundings.

3. **Mitigation Measures:** When wind turbines are proposed near historical sites, developers and stakeholders can implement mitigation measures to preserve the cultural heritage. For example, turbines can be designed to blend with the landscape, incorporating architectural elements that mirror the traditional vernacular style. Additionally, utilizing advanced technologies to reduce the visual impact, such as camouflaging techniques or innovative placement strategies, can help minimize the turbines' intrusion.

Case Study: Preservation of Stonehenge

Stonehenge, the iconic prehistoric monument in England, serves as a compelling case study for balancing cultural heritage preservation with renewable energy initiatives. The UK government has set ambitious targets for renewable energy generation, including wind power.

However, proposed wind farms near Stonehenge faced strong opposition from heritage conservationists. They argued that the visual impact of wind turbines would compromise the integrity of the site and detract from the spiritual and historical significance it holds.

To address these concerns, extensive consultations were held with heritage experts, local communities, and environmental organizations. Alternate sites for wind farms were explored, taking into account the need for clean energy and the preservation of Stonehenge's exceptional cultural value.

Ultimately, a compromise was reached, and the wind farm was relocated to a less visually intrusive location, while still contributing to renewable energy goals. This case study highlights the importance of balancing renewable energy demands with the preservation of culturally significant sites.

Conclusion

Cultural and historical preservation should be a vital consideration in the development of wind energy projects. While the transition to renewable energy is crucial, it must not come at the expense of our cultural heritage. Through thorough assessments, meaningful engagement with stakeholders, and innovative mitigation strategies, we can strike a balance that respects both clean energy goals and the preservation of our shared human history. By making conscious and informed decisions, we can ensure a sustainable future without sacrificing our cultural and historical treasures.

Economic Considerations of Wind Turbines

Cost-Benefit Analysis of Wind Turbines

Initial Installation Costs

When it comes to wind turbines, one of the biggest considerations is the initial installation costs. These costs encompass everything from purchasing the equipment to preparing the site for installation. Let's dive into the details and uncover the expenses involved in setting up wind turbines.

Purchasing Turbines

The first significant cost factor is the price of the wind turbines themselves. The cost of a turbine depends on its size, capacity, and the manufacturer. Turbines vary in size, with larger turbines generally being more expensive due to their higher power output. On average, a typical wind turbine can cost anywhere from $1.3 million to $2.2 million per megawatt (MW) of capacity. This means a 2 MW turbine can cost between $2.6 million and $4.4 million.

Although the initial investment may seem substantial, it's important to consider the long-term benefits of wind energy. Once operational, wind turbines have relatively low operating costs, making them cost-effective over their lifespan.

Site Preparation

Before installation can take place, the site needs to be prepared. This involves clearing the land, building access roads, and creating a foundation for the turbines. Site preparation costs can vary depending on the terrain, existing infrastructure, and environmental considerations.

Clearing the land and constructing access roads can add up to significant expenses, especially if the site is hilly or heavily forested. Additionally, if the site is located in a remote or hard-to-reach location, transportation costs for materials and equipment may increase.

The foundation for wind turbines can be either a concrete pad or a deep-cast foundation. Concrete pads are less expensive but may not be suitable for larger turbines or areas with high wind speeds. Deep-cast foundations, on the other hand, are more expensive but offer better stability and can accommodate larger turbines. The choice of foundation depends on various site-specific factors.

Infrastructure and Grid Connection

Another cost to consider is the infrastructure required to connect the wind turbines to the electrical grid. This includes transmission lines, transformers, and substations. The distance from the wind farm to existing grid infrastructure can significantly affect these costs.

If the wind farm is located far from the grid, the cost of building transmission lines to connect the turbines to the grid can be substantial. These costs must be balanced against the potential benefits of wind energy in that specific location.

Permitting and Legal Costs

Obtaining the necessary permits and complying with legal requirements is an important step in the installation process. Depending on the jurisdiction, wind energy projects may require permits for environmental impact assessments, land use, wildlife protection, noise compliance, and more.

The costs associated with permitting and legal compliance can vary widely depending on the location and the specific requirements of the project. Engaging with experts in environmental and legal matters can help navigate these complexities and ensure smooth project execution.

Additional Costs

In addition to the major cost factors mentioned above, there may be other smaller expenses. These can include project development and management costs, insurance fees, warranty and maintenance contracts, and the cost of any required studies or assessments.

Conclusion

The initial installation costs of wind turbines involve several factors, including the purchase of turbines, site preparation, infrastructure and grid connection, and permits and legal requirements. While these costs can be substantial, it's important to consider the long-term benefits of wind energy in terms of low operating costs and environmental sustainability. As technology advances, the initial costs are expected to decrease, making wind energy an even more viable and cost-effective solution in the future.

Maintenance and Operational Costs

Maintaining and operating wind turbines is a crucial aspect of their overall economic feasibility and sustainability. While wind energy is touted as a cost-effective renewable energy source, the ongoing expenses associated with maintenance and operation can have a significant impact on the overall financial viability of wind turbine projects. In this section, we will explore the various factors that contribute to these costs and discuss strategies for minimizing them.

Routine Maintenance

Routine maintenance is essential to ensure the optimal performance and longevity of wind turbines. It involves regular inspections, cleaning, and minor repairs. These activities are typically carried out by trained technicians who climb the towers to access the turbine components.

The frequency and extent of routine maintenance depend on several factors, including the turbine's age, location, and manufacturer's recommendations. On average, it is recommended to conduct routine inspections every six months to identify and address any potential issues before they become major problems.

Routine maintenance costs can include the following:

- Labor costs: Technicians' salaries and benefits should be factored into the maintenance budget. It is important to ensure that an adequate number of qualified personnel are available to perform inspections and repairs.

- Equipment and tools: Specialized equipment, such as cranes, lifts, and safety gear, is required for maintenance activities. These costs should be accounted for in the overall maintenance budget.

- Spare parts: Components that are subject to wear and tear, such as bearings, gearboxes, and blades, may need to be replaced periodically. Maintaining an inventory of spare parts is necessary to minimize downtime.

Corrective Maintenance

Despite regular maintenance, wind turbines may experience unexpected failures or breakdowns. Corrective maintenance refers to the repair and replacement of faulty components to restore the turbine's functionality.

The costs associated with corrective maintenance can vary widely depending on the nature of the problem and the extent of the damage. Major repairs that require component replacement or specialized expertise can be particularly expensive.

To mitigate the financial impact of corrective maintenance, it is advisable to have warranty agreements in place with turbine manufacturers or third-party service providers. These agreements can cover the cost of repairs and replacements for a specified period, allowing operators to plan and budget accordingly.

Remote Monitoring and Diagnostics

Advancements in technology have facilitated remote monitoring and diagnostics of wind turbines. Through the use of sensors and data analytics, operators can monitor the performance of their turbines in real-time and identify potential issues remotely.

Implementing remote monitoring systems can offer several benefits, including:

- Early detection of faults: By continuously monitoring turbine performance data, operators can identify deviations from normal operating conditions and take proactive measures to address potential problems before they escalate.

- Reduced maintenance costs: Remote diagnostics can help technicians diagnose issues more accurately, enabling them to come equipped with the necessary tools and spare parts. This reduces the need for multiple site visits and lowers overall maintenance costs.

- Improved turbine availability: Timely detection and resolution of faults result in reduced downtime, ensuring that the turbines are operational and generating electricity for longer periods.

While implementing remote monitoring systems incurs initial costs, the long-term benefits can outweigh the investment, leading to improved operational efficiencies and cost savings.

Staff Training and Safety Measures

To ensure effective maintenance and operational practices, it is crucial to invest in staff training and implement appropriate safety measures.

Training programs for maintenance technicians should cover a range of topics, including:

- Safety procedures: Working at heights and handling specialized equipment require proper training to minimize the risk of accidents.

- Component-specific knowledge: Technicians should be well-versed in the operation and maintenance requirements of different turbine components to carry out their tasks efficiently.

- Troubleshooting skills: A thorough understanding of turbine systems and the ability to diagnose and resolve issues are essential for minimizing downtime.

Additionally, strict adherence to safety protocols, such as wearing personal protective equipment and following safety guidelines, is paramount to ensure the well-being of maintenance personnel.

Financial Planning and Risk Management

Given the long operational life of wind turbines (typically around 20-25 years), it is essential to incorporate long-term financial planning and risk management strategies into maintenance and operational cost considerations.

Reserve Funds Setting aside reserve funds throughout the project lifespan can help cover unexpected repairs, major component replacements, and potential technological advancements that may require updates or retrofits. These funds act as a safety net, minimizing financial strain when significant expenditures arise.

Insurance Wind turbine operators often opt for insurance coverage to mitigate financial risks associated with damage, loss, or liability. Insurance policies can provide financial protection against unforeseen events such as severe weather, accidents, and equipment failures.

Performance Guarantees When purchasing wind turbines, operators can negotiate performance guarantees with manufacturers, ensuring that the turbines meet specific energy production targets. These guarantees hold the manufacturers accountable for any underperformance and may include compensation provisions.

Lifecycle Assessment Conducting a comprehensive lifecycle assessment can help determine the optimal maintenance strategies and anticipate future costs associated with component replacements and technological advancements. This assessment enables operators to plan and budget for the expected lifecycle of the turbines accurately.

Conclusion

Maintenance and operational costs are significant considerations in the deployment of wind turbine projects. Understanding the various factors contributing to these costs and implementing measures to minimize them is crucial for the long-term viability of wind energy. By investing in routine maintenance, remote monitoring, staff training, and risk management strategies, wind turbine operators can ensure optimal performance while effectively managing the financial aspects of their operations.

Cost of Intermittency and Battery Storage

One of the major drawbacks of wind turbines is their intermittent nature. Wind speeds vary throughout the day and across different seasons, resulting in fluctuations in power generation. This intermittency poses challenges for the reliable and stable supply of electricity. To address this issue, battery storage systems have emerged as a potential solution. In this section, we will explore the cost implications of intermittency and the integration of battery storage for wind energy.

Understanding Intermittency

Intermittency refers to the unpredictable nature of wind power generation. When the wind is not blowing, or its speed is below the cut-in speed of the turbines, no electricity can be generated. Conversely, when the wind speed exceeds the rated wind speed, turbines must be shut down to prevent damage.

These fluctuations in wind power output pose challenges for grid operators who need to maintain a stable power supply. Intermittent power generation not only affects the reliability of the grid but also leads to imbalances in supply and demand, which can result in voltage and frequency fluctuations.

The Role of Battery Storage

Battery storage systems provide a means to store excess electricity generated during high wind periods and release it during low wind periods. They act as a buffer, smoothing out generation fluctuations and improving grid stability.

When wind turbines generate more electricity than is immediately needed, the excess electricity can be stored in batteries. During periods of low wind, the stored electricity can be discharged to meet electricity demand, reducing the need for backup power sources or reliance on the grid. This helps in managing the intermittency of wind power and provides a more consistent supply of electricity.

Battery Storage Technologies

Several battery technologies are suitable for storing wind energy, each with its advantages and cost considerations. Lithium-ion batteries, for example, are widely used due to their high energy density, efficiency, and long cycle life. They are commonly found in electric vehicles and are increasingly being deployed for stationary energy storage applications.

Other battery technologies include flow batteries, which offer scalability and long cycle life, and sodium-sulfur batteries, known for their high energy density and ability to discharge for extended periods. Each technology has its unique characteristics and cost profiles, making it important to carefully evaluate the best fit for a specific application.

Cost Considerations

The cost of integrating battery storage with wind turbines involves various factors, including capital costs, operation and maintenance expenses, and the cost of energy storage capacity.

Capital costs encompass the construction and installation of battery storage systems. These costs vary depending on the technology used, the size of the system, and the specific project requirements. However, with advancements in technology and economies of scale, the cost of battery storage has been steadily declining in recent years.

Operation and maintenance expenses are another important consideration. Regular maintenance and monitoring are necessary to ensure the optimal performance and longevity of battery storage systems. These expenses include personnel costs, system upgrades, and periodic battery replacements.

The cost of energy storage capacity refers to the price per unit of energy storage. This cost is influenced by factors such as battery type, capacity, efficiency, and lifespan. As technology advances and demand for battery storage increases, economies of scale are expected to further drive down the cost per unit of energy storage.

Economic Benefits

Despite the additional costs associated with battery storage, there are several economic benefits to consider. First and foremost, integrating battery storage enables a more reliable and stable electricity supply, reducing the risk of power outages and disruptions.

Battery storage also facilitates the integration of higher levels of wind energy into the grid. By mitigating intermittency, it becomes possible to maximize the utilization of wind resources, ultimately reducing reliance on fossil fuel-based power generation. This transition to cleaner and renewable energy sources aligns with global efforts to combat climate change.

Moreover, the deployment of battery storage creates new job opportunities in manufacturing, installation, operation, and maintenance. The growth of the battery storage industry contributes to local and national economies, stimulating economic activity and driving innovation.

Case Study: Hornsdale Power Reserve

A notable example of the successful integration of battery storage with wind turbines is the Hornsdale Power Reserve in South Australia. It is one of the largest battery installations in the world and has been instrumental in stabilizing the region's electricity grid.

The Hornsdale Power Reserve, consisting of Tesla's lithium-ion battery technology, provides grid services such as frequency regulation and rapid response to demand fluctuations. It has improved grid stability and reduced the reliance on traditional spinning reserve power plants.

The success of the Hornsdale Power Reserve demonstrates the potential of battery storage in enhancing the reliability and efficiency of wind energy systems. It serves as a testament to the economic viability and operational benefits of integrating battery storage with wind turbines.

Conclusion

The cost of intermittency in wind turbines can be mitigated through the integration of battery storage systems. While there are additional costs associated with the construction, operation, and maintenance of these systems, the economic benefits are substantial.

Battery storage enhances the reliability and stability of wind energy systems, allowing for a more consistent supply of electricity. It enables a higher penetration of wind energy into the grid, reducing reliance on fossil fuels and contributing to environmental sustainability. Moreover, the deployment of battery storage drives economic growth and job creation.

As battery technologies continue to advance and costs further decrease, the integration of battery storage with wind turbines will become an increasingly viable and attractive option for a reliable and sustainable energy future.

Job Creation and Economic Stimulus

In addition to its environmental benefits, the development of wind energy infrastructure brings significant economic advantages. One of the key factors that contributes to the growth of the wind industry is its ability to create jobs and stimulate local economies. In this section, we will explore the various aspects of job creation and economic stimulus associated with wind turbines.

Job Creation

The wind energy sector has seen remarkable growth in recent years, leading to a surge in job opportunities. The installation, operation, and maintenance of wind turbines require a diverse range of skills, offering employment opportunities in various sectors.

1. Skilled Jobs: The installation of wind turbines necessitates the expertise of engineers, electricians, and technicians. These skilled professionals conduct site assessments, design the layout of wind farms, install turbines, and ensure their efficient operation. They play a crucial role in the development and growth of the wind energy sector.

2. Manufacturing: Wind turbine components, such as blades, towers, generators, and control systems, are manufactured in specialized factories. The manufacturing sector not only provides a substantial number of jobs but also drives innovation through research and development, leading to improvements in turbine technology.

3. Supply Chain: The wind energy industry relies on a complex supply chain, involving raw materials suppliers, logistics providers, and various services related to equipment transportation, installation, and maintenance. This intricate network generates employment opportunities across multiple sectors.

4. Operations and Maintenance: Wind turbines require regular maintenance and monitoring to ensure optimal performance. This ongoing work includes inspection, repair, and troubleshooting, which create a demand for skilled technicians and support staff. The operation and maintenance phase of wind farms provides long-term employment.

5. Support Services: The wind energy sector generates jobs in supporting services such as project management, finance, legal services, and consulting. These positions contribute to the overall growth and stability of the industry.

By fostering job creation across various sectors, the wind energy industry provides a strong foundation for economic development and enhances the livelihoods of local communities.

Economic Stimulus

Apart from job creation, wind turbine installations and the subsequent operation of wind farms stimulate economic growth in several ways.

1. Local Purchasing: Wind energy projects often prioritize using local resources for construction and operation. This leads to increased demand for construction materials, heavy machinery rentals, maintenance equipment, and other goods and services from local suppliers. This boosts the local economy by keeping money within the community and creating a multiplier effect.

2. Tax Revenue: The establishment of wind farms contributes to tax revenues at various levels. Local governments receive property taxes and other levies, which can be used to invest in public infrastructure, healthcare, education, and other social services. This revenue source helps address community needs and improve overall quality of life.

3. Increased Spending: Wind farm development attracts workers who spend their wages on local goods and services, stimulating the retail and hospitality sectors. Additionally, the influx of visitors to wind farms who come for tourism or educational purposes also boosts local businesses such as hotels, restaurants, and recreational activities.

4. Economic Diversification: Wind energy projects offer opportunities for regions to diversify their economies. Areas that traditionally relied on industries such as coal mining or manufacturing can transition to renewable energy production, reducing dependency on a single sector and creating more stable and sustainable economic growth.

5. Community Benefits: Wind energy companies often negotiate agreements with local communities to provide additional benefits, such as community

development funds or profit-sharing arrangements. These benefits help support local initiatives, infrastructure, and social programs.

By providing job opportunities, generating tax revenue, increasing local spending, and promoting economic diversification, wind energy projects act as a catalyst for sustainable economic growth and provide long-term economic benefits for communities.

Overall, wind turbines not only contribute to a cleaner and more sustainable future, but they also play a vital role in job creation and stimulating local economies. The growth of the wind energy sector presents significant opportunities for professional development and economic advancement in various industries and regions.

Subsidies and Government Incentives

In the world of wind energy, subsidies and government incentives play a crucial role in promoting the growth and development of wind turbines. These financial mechanisms are designed to make wind power more competitive with traditional sources of energy and stimulate investment in renewable energy projects. Let's take a closer look at the various types of subsidies and incentives available, their importance, and the controversies surrounding them.

Feed-in Tariffs

One of the most widely used mechanisms for supporting wind energy is the feed-in tariff (FiT) system. Under a FiT, renewable energy producers are guaranteed a fixed premium price for the electricity they generate over a long-term contract. This price is often higher than the current market rate, providing an incentive for wind power developers to invest in new projects.

The main advantage of FiTs is the stability and predictability they offer to project developers. By ensuring a fixed revenue stream, FiTs lower the investment risk associated with wind energy projects, making them more attractive to investors. As a result, FiTs have played a vital role in kickstarting the growth of the wind energy sector in many countries.

However, FiTs have faced criticism for potentially driving up electricity prices for consumers. Critics argue that the higher costs associated with guaranteed premium prices are ultimately passed on to electricity consumers, putting an additional burden on households and businesses. The debate surrounding the fairness of FiTs is an ongoing one, with proponents highlighting the long-term benefits of renewable energy and opponents emphasizing the short-term costs.

Production Tax Credits

Production tax credits (PTCs) are another form of subsidy commonly used to incentivize wind energy production. Under the PTC system, wind energy producers receive a tax credit for each kilowatt-hour of electricity they generate. This financial incentive directly reduces the tax liability of wind power developers, effectively lowering the cost of wind energy projects and making them more financially viable.

PTCs have been instrumental in encouraging wind power development, particularly in the United States. By reducing the upfront costs of wind projects and providing a stable revenue stream, PTCs have attracted investment and contributed to the rapid expansion of wind energy capacity in the country.

However, PTCs have also faced criticism for their inconsistent and uncertain nature. The intermittent renewal of PTCs, with sometimes short-term extensions, creates uncertainty for wind power developers and investors. This unpredictability can hinder long-term planning and slow down the growth of the wind industry. Balancing the need for stable government support with the concerns of taxpayers and policymakers remains a challenge in implementing PTCs effectively.

Investment Grants and Subsidies

In addition to feed-in tariffs and production tax credits, investment grants and subsidies are often provided by governments to promote wind energy projects. These financial incentives can take various forms, such as direct grants, low-interest loans, or tax incentives that allow for depreciation of wind power assets over time.

Investment grants and subsidies help to offset the high initial costs associated with wind energy projects. By making capital more accessible and reducing the financial burden on developers, these incentives encourage investment in renewable energy and facilitate the deployment of wind turbines.

Controversies surrounding investment grants and subsidies mainly revolve around the allocation of public funds. Critics argue that governments should focus on providing support to technologies that are on the brink of commercial viability rather than subsidizing mature technologies like wind power. Additionally, the effectiveness of grants and subsidies in achieving long-term sustainability goals and creating lasting impacts is an ongoing debate.

Importance of Subsidies and Government Incentives

Subsidies and government incentives play a vital role in driving the growth of wind energy by making it economically viable and attractive to investors. They help bridge the gap between the higher costs associated with renewable energy and the lower costs of traditional fossil fuel-based energy generation.

By providing stable and predictable financial support, subsidies and incentives reduce the investment risk for wind power developers. This, in turn, encourages private investment, stimulates job creation, and promotes research and development in wind energy technology. Moreover, the growth of wind energy contributes to reducing greenhouse gas emissions, mitigating climate change, and securing a sustainable energy future.

Caveats and Considerations

While subsidies and government incentives have been instrumental in the expansion of wind power, it is important to examine their long-term implications and potential drawbacks. Here are a few caveats and considerations to keep in mind:

- **Market Distortion:** Subsidies can distort the market and create an artificial advantage for wind energy, potentially crowding out other renewable energy technologies. Striking a balance between supporting wind power and fostering a diverse and competitive renewable energy sector is crucial.

- **Budgetary Constraints:** Government support for wind energy is subject to budgetary limitations. As the cost of subsidies and incentives increases, policymakers must evaluate their affordability and consider the optimal allocation of public funds.

- **Technology Advancements:** As wind energy technology advances and costs decline, the reliance on subsidies may decrease. A phased reduction in subsidies, coupled with ongoing research and development efforts, can drive innovation and foster cost-competitive wind power.

- **Regional Considerations:** The effectiveness of subsidies and incentives can vary depending on the region's unique circumstances. Factors like wind resources, grid infrastructure, and local energy policies should be taken into account to ensure targeted and efficient support.

In conclusion, subsidies and government incentives have played a crucial role in promoting wind energy by reducing financial barriers, attracting investment, and stimulating the growth of the wind industry. Despite the debates and controversies surrounding these financial mechanisms, their importance in transitioning to a sustainable energy future cannot be ignored. A well-designed and balanced approach to subsidies and incentives is essential to maximize the benefits of wind power while addressing concerns related to cost, market distortion, and budgetary limitations.

Impact on Property Values

Property Devaluation near Wind Farms

Property devaluation near wind farms is a significant concern for homeowners and investors alike. The presence of wind turbines in close proximity to residential areas has the potential to impact property values. In this section, we will explore the factors contributing to property devaluation and discuss the possible solutions and mitigations.

Understanding Property Values

Before delving into the specific impacts of wind farms on property values, it is essential to understand the factors affecting property valuations in general. Several key factors influence property values, including location, amenities, economic conditions, and market demand.

Location plays a vital role in determining property values. Homes situated in desirable neighborhoods, close to schools, parks, and other conveniences, tend to command higher prices. Conversely, properties located near undesirable features like busy highways, industrial sites, or waste treatment facilities may experience lower valuations.

Amenities such as shopping centers, recreational facilities, and access to public transportation can also impact property values positively. Additionally, economic conditions, such as job opportunities and the overall health of the local economy, influence property valuations.

Market demand is another crucial factor affecting property values. Higher demand drives competition among buyers, leading to increased prices. On the other hand, lower demand can result in reduced property values.

Factors Contributing to Property Devaluation near Wind Farms

While wind farms offer numerous benefits in terms of renewable energy generation, their presence near residential areas can lead to property devaluation. Several factors contribute to this phenomenon:

IMPACT ON PROPERTY VALUES

1. Visual Impact: The visual impact of wind turbines is one of the primary concerns for homeowners. Some may find the sight of towering turbines obstructing their views unappealing, leading to a perceived reduction in property value. The aesthetic preferences of potential buyers can vary, making it challenging to quantify the exact impact on property values.

2. Noise Pollution: Wind turbines generate noise during operation, which can be a significant nuisance for nearby residents. The constant hum or swooshing sound of rotating blades may disturb the peace and quiet that homeowners value, degrading their quality of life. This noise pollution can negatively impact property values.

3. Flicker Effect: The rotation of wind turbine blades can cause a flickering effect, casting shadows over neighboring properties. This intermittent change in lighting conditions can be irritating for residents and affect the desirability of homes in the vicinity. Consequently, property values may decrease as potential buyers may be deterred by this flickering effect.

4. Fear of Health Effects: While scientific studies have not established a direct link between wind turbines and adverse health effects, some individuals may have concerns about potential health risks. These fears may stem from misinformation or the psychological impact of living near large structures. The perception of health risks can influence property values.

5. Lack of Privacy: Wind farms often require large areas of land, resulting in decreased privacy for neighboring properties. Homeowners may feel uneasy about the loss of privacy and the constant presence of wind farm personnel conducting maintenance and inspections. This loss of privacy can contribute to property devaluation.

Mitigating Property Value Impacts

Managing the impacts of wind farms on property values requires a balanced approach that considers the concerns of homeowners and the need for renewable energy. Here are some strategies to mitigate property value impacts:

1. Set Appropriate Wind Farm Setback Distances: Establishing adequate setback distances between wind turbines and residential areas can help mitigate the visual impact, noise pollution, and flicker effects. This distance should take into account local topography, prevailing wind patterns, and the size and height of the turbines.

2. Implement Noise Regulations: Implementing strict noise regulations that consider both the decibel levels and the duration of exposure can help mitigate noise pollution near residential areas. By ensuring wind turbines operate within acceptable noise limits, the negative impact on property values can be minimized.

3. Utilize Visual Screening Techniques: Employing visual screening techniques, such as landscaping or natural buffers, can help mitigate the visual impact of wind turbines. Strategic placement of trees, shrubs, and other vegetation can create a visually appealing barrier between the turbines and residential properties, reducing the perceived negative impact.

4. Educate the Public and Address Health Concerns: Public education campaigns aimed at providing accurate information about wind energy and dispelling misconceptions can help alleviate fears and concerns related to health effects. Engaging with the community and addressing their concerns through transparent communication channels can contribute to maintaining property values.

5. Compensation and Incentive Programs: Developing compensation and incentive programs for homeowners living near wind farms can help offset any property value impacts. These programs can include financial compensation, reduced energy costs, or community benefits, ensuring that affected homeowners feel adequately compensated for any perceived loss in property value.

Case Study: Falmouth, Massachusetts

The town of Falmouth, Massachusetts, provides an insightful case study on the impacts of wind turbines on property values. Falmouth installed two wind turbines, known as Wind 1 and Wind 2, in 2010. However, shortly after their installation, nearby residents began reporting adverse effects on their health and quality of life.

Multiple lawsuits were filed, citing the negative impact on property values and the noise and flicker effects caused by the turbines. As a result, Wind 1 was ordered to shut down, and Wind 2 faced similar legal challenges.

The Falmouth case highlights the importance of proper planning, setback distances, and addressing community concerns before the installation of wind turbines. Failing to account for these factors can have severe consequences, including legal battles and property devaluation.

Conclusion

Property devaluation near wind farms is a complex issue that requires careful consideration of various factors. While the presence of wind turbines can impact property values, implementing appropriate setback distances, noise regulations, visual screening techniques, and compensation programs can help mitigate these effects. Public education and addressing health concerns also play a vital role in maintaining property values. By striking a balance between renewable energy goals and preserving property values, we can ensure a sustainable future for both homeowners and the environment.

Legal issues and property rights

The development of wind turbines, although considered an important step towards renewable energy, has raised several legal and property rights issues. These concerns primarily revolve around the impact of wind farms on neighboring property values, the potential encroachment on landowner rights, and the need for compensation and mitigation for affected homeowners.

Property devaluation near wind farms

One of the main concerns raised by homeowners near wind farms is the potential devaluation of their properties. The presence of wind turbines, especially in close proximity, can have a negative impact on the aesthetics and scenic value of the landscape. This visual pollution can deter potential buyers and reduce the demand for properties, leading to a decrease in property values.

Several studies have examined the effect of wind farms on property prices. While some studies suggest no significant impact, others have found evidence of property devaluation. The extent of devaluation depends on various factors such as distance from the wind farm, size and number of turbines, visual impact, and the perception of noise.

Legal issues and property rights

The development of wind turbines involves legal and regulatory considerations to ensure that landowner rights are respected and that any potential harm is adequately addressed. Some of the key legal issues and property rights related to wind turbines include:

1. **Zoning and land-use regulations:** Local zoning and land-use regulations play a vital role in determining where wind turbines can be installed. These

regulations define the permitted locations for wind farms and specify setbacks from property lines, roads, and residential areas. They aim to balance the interests of wind energy development with the protection of surrounding land uses and property values.

2. **Lease agreements:** Developers typically enter into lease agreements with landowners to secure the right to install wind turbines on their properties. These agreements outline the payments, terms, and conditions for the use of the land. Landowners should carefully review the terms to ensure they are fair and account for any potential impacts on their property and quality of life.

3. **Eminent domain:** In some cases, wind energy developers may invoke eminent domain, a legal process that allows the government or private entities to acquire private property for public use. This power is controversial, and its use for wind energy projects has faced legal challenges in some jurisdictions. It is essential for affected landowners to be aware of their rights and seek legal representation if necessary.

4. **Nuisance claims:** Homeowners living near wind farms may file nuisance claims if they can demonstrate that the turbines' operation substantially interferes with the use and enjoyment of their property. Such claims may encompass issues such as noise, flicker effect, or vibrations caused by the turbines. Courts often assess the reasonableness of the alleged nuisance, taking into account local regulations, industry standards, and the impact on property values.

5. **Compensation and mitigation:** To address property devaluation and potential impacts on neighboring landowners, compensation and mitigation measures may be required. Developers may be responsible for compensating affected homeowners for any loss in property value. Additionally, mitigation measures such as sound barriers, visual screening, or relocation of turbines may be implemented to minimize adverse impacts.

Balancing interests and finding solutions

The legal issues and property rights surrounding wind turbines require a delicate balance between promoting renewable energy and protecting the rights of homeowners. Clear and transparent regulations, fair lease agreements, and effective mitigation measures can help resolve conflicts and mitigate the negative impacts on property values and quality of life.

Moreover, engaging the community in the decision-making process and providing avenues for public participation can foster understanding and address concerns. Dialogue between developers, homeowners, and regulators is crucial to finding mutually acceptable solutions that strike a balance between sustainable energy generation and the preservation of property rights.

Case study: Property rights in Texas

Texas, with its vast open spaces and abundant wind resources, has become a major hub for wind energy development. The state has implemented legislation to protect property rights while promoting wind power. The Texas Landowner's Bill of Rights ensures that landowners are fully informed about their rights when negotiating with wind developers. It provides guidelines on lease agreements, compensation, and the legal recourse available to resolve disputes.

Despite these measures, challenges remain. In some cases, wind farm developers have faced opposition from landowners concerned about the impact on their property values and the environmental consequences of large-scale wind energy projects. Legal battles have ensued, highlighting the complex interplay between private property rights and the collective interests of renewable energy development.

To navigate these challenges, the state of Texas encourages open communication and negotiation between developers and landowners. Landowners are advised to seek legal counsel to ensure that their property rights are protected, and developers are encouraged to address concerns and provide fair compensation packages to affected parties.

Conclusion

Legal issues and property rights are significant considerations in the development of wind turbines. Building wind farms that are sensitive to landowner rights, property values, and environmental concerns is crucial for the success of renewable energy initiatives. Balancing the benefits of clean energy with the rights of homeowners requires clear regulations, fair lease agreements, and transparent processes that encourage open dialogue and participation. By addressing legal and property rights issues effectively, society can maximize the potential of wind energy while minimizing any adverse impacts on individuals and communities.

Compensation and mitigation for affected homeowners

The installation of wind turbines near residential areas can have a significant impact on homeowners. It can lead to decreased property values, noise pollution, and visual disturbances. In order to address these concerns and ensure proper compensation and mitigation for affected homeowners, it is necessary to have appropriate policies and regulations in place.

1. **Property devaluation near wind farms:** One of the major concerns for homeowners living near wind farms is the potential devaluation of their properties. Studies have shown that the proximity of wind turbines can have a negative impact on property values. To address this issue, it is important to establish a fair and transparent system for compensating homeowners for any losses in property value.

2. **Legal issues and property rights:** Another aspect to consider is the legal framework surrounding wind turbine installations and property rights. Homeowners should have the right to object to the construction of wind turbines near their properties if they can demonstrate valid reasons such as potential health risks or significant aesthetic impact. Clear guidelines should be established to protect the property rights of homeowners and ensure their voices are heard.

3. **Compensation for affected homeowners:** Compensation should be provided to homeowners who experience a decline in property value or other negative impacts due to the presence of wind turbines. The compensation should be fair and take into account the extent of the impact and the financial losses incurred. This could include monetary compensation or other measures such as property tax reductions.

4. **Mitigation measures:** In addition to compensation, it is essential to implement mitigation measures to minimize the impact of wind turbines on homeowners. This could involve the use of effective noise-reducing technologies to address noise pollution concerns or the implementation of landscaping strategies to improve visual aesthetics. By implementing appropriate mitigation measures, the negative effects on homeowners can be mitigated, and their quality of life can be preserved.

5. **Community engagement and involvement:** To ensure that homeowners' concerns are addressed effectively, it is crucial to involve them in the decision-making process. Open and transparent communication channels should be established, allowing homeowners to express their concerns and participate in discussions regarding wind turbine installations. This can help build trust and ensure that their needs are taken into account when developing policies and regulations.

6. **Education and awareness programs:** Lastly, education and awareness programs should be implemented to inform homeowners about the benefits and

drawbacks of wind turbines. Providing accurate information about the potential impacts, compensation options, and mitigation measures can help alleviate fears and misconceptions. These programs can empower homeowners to make informed decisions and engage in constructive dialogue with relevant authorities.

In conclusion, addressing the concerns of homeowners affected by wind turbines requires a comprehensive approach. It involves fair compensation for property devaluation, protection of property rights, implementation of appropriate mitigation measures, community engagement, and education programs. By adopting these measures, we can strive for a balance between renewable energy generation and the well-being of homeowners.

Technical Limitations and Challenges of Wind Turbines

Wind Variability and Unpredictability

Reliability and Grid Integration Challenges

Reliability and grid integration are crucial aspects when considering the deployment of wind turbines. While wind energy is a promising source of renewable power, there are certain challenges that need to be addressed to ensure its reliable integration into the existing grid infrastructure.

Intermittency and Power Variability

One of the primary challenges associated with wind energy is its inherent intermittency and power variability. The amount of power generated by wind turbines is dependent on the speed and consistency of the wind. As the wind is a natural resource, it is subject to fluctuations and cannot be controlled.

This intermittency poses challenges for grid operators who need to maintain a stable supply-demand balance at all times. Without proper management, sudden changes in wind power output can lead to grid instability and potential blackouts. To address this challenge, grid operators employ various strategies such as forecasting, backup power sources, and energy storage technologies.

Grid Balancing and Flexibility

The integration of large-scale wind farms into the grid requires careful balancing of electricity supply and demand. Unlike traditional power plants, which can be dispatched and ramped up or down as needed, the output of wind turbines is determined by the availability of wind resources.

Grid balancing becomes more challenging as the share of wind energy in the grid increases. Power systems need to be flexible enough to accommodate the variability of wind power and respond to rapid changes in generation. This flexibility can be achieved through the use of advanced grid management techniques, such as demand response programs, smart grids, and grid-scale energy storage.

Transmission Infrastructure

Another challenge in the grid integration of wind turbines is the need to upgrade and expand the transmission infrastructure. Wind resources are often located in remote areas with limited existing transmission capacity. Connecting wind farms to the grid requires the construction of new transmission lines, which can be costly and time-consuming.

Furthermore, the transmission of electricity over long distances from wind-rich regions to population centers may result in transmission losses. To minimize these losses, high-voltage direct current (HVDC) transmission systems are often utilized. HVDC technology allows for efficient long-distance transmission and enables the integration of remote wind farms into the grid.

Voltage and Frequency Control

Voltage and frequency control are critical for maintaining grid stability and ensuring the reliable operation of electrical devices. The integration of variable wind power can introduce challenges in maintaining stable voltage and frequency levels.

When wind power generation exceeds the local demand, the excess electricity needs to be exported to the grid. This can cause voltage and frequency fluctuations if not properly managed. Advanced control systems, such as power electronics-based solutions, are used to regulate voltage and frequency and maintain grid stability.

Grid Planning and System Operation

Proper grid planning and system operation are essential for the successful integration of wind energy. The location and capacity of wind farms need to be carefully determined to optimize the utilization of wind resources and minimize transmission losses. Additionally, the operation of wind turbines must be coordinated with other power generation sources to ensure grid stability and reliability.

Sophisticated modeling and simulation tools are utilized to assess the impact of wind power on the grid and to optimize its integration. These tools aid in predicting the behavior of the entire power system under different operating conditions and help in making informed decisions regarding the deployment and operation of wind turbines.

Example: The Great Texas Blackout of 2021

An unfortunate example of the challenges associated with grid integration and reliability issues occurred in Texas in February 2021. Severe winter weather conditions led to a significant increase in electricity demand for heating, while at the same time, many wind turbines froze due to unpreparedness for such extreme conditions.

This combination of events resulted in a sudden and substantial decrease in the available wind power generation capacity, exacerbating the strain on the grid. The unanticipated loss of wind power, coupled with other issues in the power system, led to widespread blackouts across the state, leaving millions of people without electricity for days.

The Texas blackout emphasizes the importance of considering the reliability and grid integration challenges associated with wind turbines. It serves as a reminder that proper planning, weather resilience, and backup solutions are crucial to ensure reliable power supply, especially during extreme weather events.

Conclusion

Reliability and grid integration challenges are significant considerations when it comes to wind energy deployment. While wind turbines offer several benefits as a source of clean and renewable power, their intermittent nature and variable output pose challenges for grid operators and system planners.

Addressing these challenges requires a combination of advanced forecasting techniques, flexible grid management strategies, robust transmission infrastructure, and efficient control systems. By overcoming these obstacles, wind energy can become a more reliable and integrated part of the overall energy mix, contributing to a greener and more sustainable future.

Intermittency and Energy Storage

Intermittency is one of the major challenges faced by wind turbines, as the wind is not constant and can fluctuate in speed and direction. This variability in wind conditions directly affects the reliability and stability of electricity generation from wind turbines. During periods of low wind speeds, wind turbines may not produce sufficient electricity to meet the demand, while during periods of high wind speeds, excess electricity may be generated, which could potentially overload the grid.

To address this issue, energy storage solutions are crucial in balancing the intermittent generation of wind turbines with the continuous demand for electricity. Energy storage allows excess electricity to be stored during periods of high wind speeds and then used during periods of low wind speeds, ensuring a steady and reliable supply of electricity.

Types of Energy Storage

There are several types of energy storage technologies that can be employed to mitigate the intermittency of wind turbines. Let's explore some of the most common ones:

1. **Battery Storage:** Batteries are an effective and widely-used energy storage option. They store electrical energy in chemical form and can release it as needed. Lithium-ion batteries, in particular, have gained popularity due to their high energy density and efficiency. Battery storage systems can be installed at wind farms or integrated into the grid infrastructure to store excess electricity generated during periods of high wind speeds and discharge it when needed.

2. **Pumped Hydroelectric Storage:** Pumped hydroelectric storage is currently the most common form of large-scale energy storage. It involves storing energy by pumping water from a lower reservoir to a higher reservoir during periods of excess electricity generation and releasing it through turbines to generate electricity when demand exceeds supply. This technology requires suitable geographical conditions, such as elevated terrains and available water sources, to create the necessary reservoirs.

3. **Compressed Air Energy Storage (CAES):** CAES is another method of storing excess electricity. During periods of high wind speeds, electricity is used to compress air and store it in underground caverns. When electricity demand increases, the compressed air is released, expanded through turbines, and converted back to electricity. CAES can provide both short-term and long-duration storage, depending on the design of the system.

4. **Flywheel Energy Storage:** Flywheel energy storage systems store energy in the form of rotational motion. When excess electricity is generated by wind turbines, it is used to accelerate a flywheel, storing the energy as kinetic energy. When demand exceeds supply, the kinetic energy is converted back

into electrical energy. Flywheel energy storage systems have fast response times and can be particularly useful in providing frequency regulation.

Challenges and Considerations

While energy storage technologies offer solutions to address the intermittency of wind turbines, there are various challenges and considerations that need to be taken into account:

1. **Cost:** Implementation and maintenance costs of energy storage systems can be substantial. The cost-effectiveness of energy storage needs to be carefully evaluated to ensure the overall viability of wind energy projects.

2. **Efficiency:** Energy storage systems have inherent energy losses during the charging and discharging processes. Maximizing the efficiency of energy storage technologies is crucial to minimize energy wastage.

3. **Scalability:** Energy storage systems should be scalable to accommodate the increasing penetration of wind energy into the grid. The capacity and capabilities of storage technologies need to align with the growing demands of the electricity grid.

4. **Environmental Impact:** The construction and operation of energy storage systems may have environmental consequences. Evaluating the environmental impact of these technologies and ensuring their sustainability is essential.

5. **Integration:** Integrating energy storage systems with existing grid infrastructure can pose technical and logistical challenges. The compatibility of storage technologies with the grid and their ability to smoothly integrate into the electricity system need to be considered.

Real-World Example: Tesla's Hornsdale Power Reserve

One notable example of a large-scale energy storage project is the Hornsdale Power Reserve in South Australia, which was developed by Tesla in collaboration with the South Australian government. The project utilizes lithium-ion battery storage to address the intermittency of wind energy in the region.

The Hornsdale Power Reserve is currently the world's largest lithium-ion battery storage installation, with a total capacity of 150 megawatts (MW) and a storage capacity of 194 megawatt-hours (MWh). This energy storage system plays a crucial role in stabilizing the electricity grid by absorbing and supplying power within milliseconds.

The project has proven highly successful in providing rapid response times during fluctuations in wind energy generation, effectively smoothing out the intermittency and improving grid stability. It has also demonstrated the commercial viability and significant potential of energy storage to support renewable energy integration.

Conclusion

Intermittency is a challenge inherent to wind turbines, and addressing it is crucial to ensure a reliable and stable electricity supply. Energy storage technologies offer effective solutions to balance the intermittent generation of wind turbines with the demand for electricity. From battery storage to pumped hydroelectric storage, various energy storage options can be employed based on geographical, economic, and technical considerations.

As the global energy landscape continues to evolve, the development and implementation of efficient and scalable energy storage systems will play a vital role in maximizing the potential of wind energy and other renewable sources. By integrating intermittent renewable generation with reliable energy storage, we can build a more sustainable and resilient energy future.

Bibliography

[1] Tesla, Inc. *Hornsdale Power Reserve.*
https://www.tesla.com/en_gb/powerpack/hornsdale.

Siting and Placement Constraints

When it comes to wind turbines, finding the perfect spot for placement is crucial. Siting and placement constraints are one of the many factors that must be considered in order to ensure the successful installation and operation of wind turbines. Let's dive into the challenges and considerations involved in this process.

Environmental Factors

The first thing to consider when siting wind turbines is the environmental impact. Wind turbines should be placed in areas where the wind flow is consistent and strong. This requires assessing wind resource potential through extensive data collection and analysis. It is important to identify locations with high wind speeds and low turbulence for optimal turbine performance.

Furthermore, the impact on wildlife and their habitats must be taken into account. Endangered species, migratory patterns, and nesting grounds can all be affected by the presence of wind turbines. Thorough environmental assessments and studies are necessary to determine the potential risks and develop mitigation strategies.

Land Use and Land Ownership

Another constraint to consider is land use and land ownership. Wind turbines require significant space for installation, with each turbine typically occupying several acres of land. Securing land leases or purchases can be a complex process that involves negotiation with landowners, local communities, and government agencies.

Moreover, the suitability of the terrain and soil conditions are crucial to ensure a stable foundation for the wind turbines. Detailed geotechnical investigations are conducted to assess the feasibility of the site and determine the appropriate foundation design.

Aesthetics and Community Acceptance

The visual impact of wind turbines on the landscape cannot be ignored. Some individuals may have concerns about the intrusion of these tall structures on scenic views or cultural/historical landmarks. Balancing the benefits of renewable energy with the preservation of aesthetics and cultural heritage requires careful consideration and community engagement.

Community acceptance is vital for the successful implementation of wind turbines. Public perception and opinion play a significant role in the approval process. Engaging with local communities, addressing their concerns, and providing clear communication about the benefits of wind energy are essential for gaining support for the project.

Grid Connection and Transmission

The proximity to existing electrical infrastructure and the capacity of the grid to handle the generated power are important considerations. Wind farms need to be located close to transmission lines to efficiently transport the electricity generated. The availability of suitable substations and connection points to the grid should be assessed during the planning phase.

Intermittency and variability of wind power generation pose additional challenges. The fluctuating nature of wind energy requires careful coordination with the grid operator to ensure grid stability and reliability. This may involve the installation of advanced forecasting systems and energy storage solutions to smooth out power output.

Legal and Regulatory Framework

Navigating through the legal and regulatory framework is another significant constraint. Wind turbine projects must comply with environmental regulations, land use policies, zoning ordinances, and local building codes. Obtaining the necessary permits and approvals can be a lengthy and complex process that requires expertise in navigating bureaucracy.

Additionally, potential conflicts with other industries, such as aviation or defense, must be addressed. Wind turbines can interfere with radar systems and pose risks to aviation navigation. Collaboration with these industries and regulatory agencies is essential to find suitable solutions and ensure safety.

Case Study: The Battle with NIMBYism

One of the most common challenges faced in siting wind turbines is NIMBYism – "Not In My Backyard" syndrome. Despite the benefits of renewable energy, some local communities may oppose wind turbine projects due to concerns about noise, visual impact, or perceived decrease in property values.

Dealing with NIMBYism requires proactive community engagement, presentations of accurate facts, and dispelling myths associated with wind energy. Collaborative workshops and informational sessions can address concerns directly and help foster a better understanding of the benefits and minimal impact of wind turbines.

Conclusion

Siting and placement constraints pose significant challenges in the installation of wind turbines. Environmental impact, land use, aesthetics, community acceptance, grid connectivity, legal frameworks, and NIMBYism all play a role in determining the success of wind projects.

Developing strategies to navigate these constraints and strike a balance between environmental considerations, community needs, and renewable energy goals is crucial. With careful planning, effective communication, and continuous engagement, wind energy can be harnessed responsibly and contribute to a cleaner, more sustainable future.

Noise and Health Concerns

Noise Pollution and Psychological Impact

Noise pollution is a problem that affects many aspects of our daily lives, from the blaring horns of rush hour traffic to the constant drone of construction work. However, the noise generated by wind turbines is a unique and often underestimated aspect of their impact. In this section, we will explore the various ways in which noise pollution from wind turbines can have a psychological impact on individuals and communities.

Understanding Noise Pollution from Wind Turbines

To fully understand the psychological impact of noise pollution from wind turbines, it is important to first grasp the nature of the noise produced by these giant machines. The primary source of noise from wind turbines comes from the rotation of the blades as they cut through the air, as well as mechanical vibrations within the structure. This creates a low-frequency noise that can be perceived as a constant hum or whooshing sound.

The noise levels generated by wind turbines can vary depending on factors such as wind speed, turbine design, and distance from the turbine. Typically, noise levels decrease as you move further away from a turbine, but even at significant distances, the audible noise can still be a source of annoyance and disturbance.

Psychological Effects of Wind Turbine Noise

The persistent presence of noise from wind turbines can have a range of psychological effects on individuals and communities, impacting their quality of life and overall well-being. These effects can manifest in various ways:

Stress and Sleep Disturbances Wind turbine noise has been found to be a significant source of stress and sleep disturbances for individuals living in close proximity to these structures. The constant hum or whooshing sound can disrupt sleep patterns, leading to insomnia and fatigue. Prolonged exposure to noise can also trigger elevated stress levels, affecting overall mental health.

Annoyance and Irritability Wind turbine noise can be highly annoying to individuals, causing irritability and frustration over time. The persistent nature of the noise can make it difficult to concentrate on tasks, leading to decreased productivity and a general sense of unease.

Decreased Quality of Life The negative psychological effects of wind turbine noise can contribute to a decreased quality of life for those living in affected areas. The constant disruption and annoyance can lead individuals to avoid outdoor activities and limit their enjoyment of their surroundings. This can have a particularly significant impact on communities that rely on tourism and outdoor recreation for economic vitality.

Mitigating the Psychological Impact

Addressing the psychological impact of wind turbine noise requires a comprehensive approach that takes into account the unique challenges posed by these structures. Here are some strategies that can be employed to mitigate the psychological impact:

Proper Siting and Setbacks Choosing appropriate locations for wind farms and implementing adequate setbacks can help minimize the impact of noise on nearby communities. By ensuring that turbines are positioned at a safe distance from residential areas, the levels of noise experienced by individuals can be reduced.

Noise Monitoring and Regulation Implementing strict noise monitoring and regulation guidelines can help ensure that wind turbines comply with acceptable noise levels. This can involve setting maximum allowable noise limits and regularly monitoring and enforcing compliance. Additionally, incorporating noise mitigation measures, such as quieter turbine designs and noise barriers, can also help reduce the psychological impact of wind turbine noise.

Community Engagement and Education Engaging with affected communities and providing them with accurate and comprehensive information about wind turbine noise can help alleviate concerns and foster a sense of control. This can involve open dialogue, public consultations, and educational campaigns to address misconceptions and provide transparent information.

Support for Affected Individuals Providing support services for individuals experiencing psychological distress due to wind turbine noise is essential. This can include mental health services, counseling, and community support programs to help individuals cope with the impact on their well-being.

Case Study: The Psychological Impact of Wind Turbines in a Rural Community

To illustrate the real-world psychological impact of wind turbines, let's consider a case study of a rural community where a wind farm was installed within close proximity to residential areas. The constant noise generated by the turbines has led to sleep disturbances, increased stress levels, and a general decline in the quality of life for many residents. Community members have reported feeling trapped in their own homes, unable to escape the relentless noise. The local tourism industry has also suffered, as visitors are deterred by the unpleasant auditory environment.

To address these issues, the community has engaged in ongoing discussions with the wind farm operators and local authorities to identify potential solutions. This has resulted in the implementation of noise monitoring systems, the introduction of setback regulations to protect residential areas, and the provision of support services to individuals experiencing psychological distress. While challenges remain, these efforts have demonstrated the importance of a collaborative approach in mitigating the psychological impact of wind turbine noise.

Conclusion

Noise pollution from wind turbines can have a significant psychological impact on individuals and communities. The constant hum or whooshing sound can contribute to stress, sleep disturbances, annoyance, and a decreased quality of life. Mitigating the psychological effects requires careful consideration of turbine location, noise regulation, community engagement, and support services for affected individuals. By addressing these concerns, we can aim to strike a balance between renewable energy generation and the well-being of those living in proximity to wind turbines.

Health Effects of Low-Frequency Noise

Low-frequency noise generated by wind turbines has been a topic of concern among communities living near wind farms. While wind turbines are generally considered to be a clean and renewable energy source, their noise emissions, particularly in the low-frequency range, have raised questions about potential health effects on nearby residents. In this section, we will explore the health implications of low-frequency noise and its impact on human well-being.

Understanding Low-Frequency Noise

Before delving into the health effects, let's first establish a clear understanding of what low-frequency noise is. Sound can be characterized by its frequency, which refers to the number of cycles a sound wave completes in one second, measured in hertz (Hz). Low-frequency noise typically encompasses frequencies below 200 Hz.

The perception of sound varies across different individuals, and while some may feel unaffected or even find certain low-frequency sounds soothing, others may experience annoyance or discomfort. It is essential to consider individual differences in sensitivity to low-frequency noise when examining its potential health effects.

Perceived Annoyance and Psychological Impact

One of the primary concerns associated with low-frequency noise from wind turbines is its potential to cause annoyance, which can lead to various psychological impacts. Annoyance refers to a feeling of displeasure or discomfort caused by external stimuli, such as noise pollution.

Research has shown that exposure to low-frequency noise can increase annoyance levels, particularly during nighttime when background noise levels are lower. Prolonged exposure to annoying sounds can contribute to stress, sleep disturbances, and reduced overall well-being.

Moreover, annoyance caused by low-frequency noise may have cascading effects on mental health, leading to symptoms such as irritability, anxiety, and decreased cognitive performance. These psychological impacts can have significant implications for individuals living near wind farms.

Physical Health Effects

In addition to psychological impacts, low-frequency noise from wind turbines may also have physical health effects. While the research in this area is still emerging, several potential health concerns have been identified.

One aspect of concern is the potential impact of low-frequency noise on cardiovascular health. Studies have suggested that prolonged exposure to low-frequency noise may contribute to increased blood pressure and heart rate, potentially leading to cardiovascular diseases such as hypertension. Further research is needed to establish a conclusive link between wind turbine noise and cardiovascular health effects.

Another area of interest is the potential effects of low-frequency noise on sleep quality. Sleep is crucial for overall health and well-being, and disruptions in sleep patterns can have detrimental effects on various aspects of life. Some studies have indicated that exposure to low-frequency noise may cause sleep disturbances, including difficulties falling asleep, frequent awakenings, and reduced sleep quality. However, more research is needed to better understand the specific mechanisms and long-term consequences of these disruptions.

Assessing and Mitigating Health Risks

To address the potential health risks associated with low-frequency noise from wind turbines, it is crucial to have effective assessment and mitigation strategies in place. The following steps can contribute to minimizing the impact on human health:

- **Regulatory Standards:** Governments and regulatory bodies should establish clear noise emission standards for wind turbines, considering the potential health implications of low-frequency noise. These standards should be regularly reviewed and updated based on the latest scientific research.

- **Site Selection and Layout:** Proper site selection and layout of wind farms can help minimize the exposure of nearby communities to excessive noise levels. Factors such as distance, topography, and prevailing wind directions should be taken into account to ensure that turbines are positioned in a way that reduces potential noise impacts.

- **Noise Mitigation Measures:** Implementing noise mitigation measures can significantly reduce the impact of low-frequency noise on nearby residents. This can include the use of quieter turbine designs, orientation adjustments, and the installation of noise barriers between the turbines and the affected communities.

- **Public Engagement and Education:** Transparent communication and public engagement concerning wind turbine projects can help address concerns and provide accurate information about the potential health effects of low-frequency noise. Educating communities about the technologies, scientific research, and mitigation efforts can help create a better understanding of the situation.

Unconventional Solution: Soundscape Design

An unconventional yet intriguing approach to addressing the health effects of low-frequency noise involves the concept of soundscape design. Soundscape design focuses on the mindful arrangement of sounds in our environment to create a positive and harmonious auditory experience.

Applying this concept to wind farms, careful consideration could be given to not only reducing the overall noise levels but also creating a soundscape that is soothing and pleasant to nearby communities. This could involve incorporating natural sounds, such as gentle wind rustling through trees or birdsong, alongside the sounds produced by wind turbines. By carefully curating the acoustic environment, the potential negative health effects of low-frequency noise may be mitigated, fostering a more positive relationship between wind farms and communities.

Conclusion

The health effects of low-frequency noise from wind turbines remain an important topic of research and discussion. While some individuals may experience annoyance or potential health impacts, it is crucial to consider a multidisciplinary approach to address these concerns.

Through comprehensive assessment, regulatory standards, site selection, noise mitigation measures, public engagement, and unconventional solutions like soundscape design, we can find a balance between renewable energy generation and the well-being of communities living near wind farms. By prioritizing scientific research, stakeholder engagement, and innovative approaches, we can create a future where the benefits of wind energy coexist harmoniously with the health and well-being of individuals.

Public Perception and Acceptance

Public perception and acceptance play a crucial role in the successful implementation of any wind turbine project. While renewable energy is generally viewed as a positive and necessary step towards a sustainable future, it is not uncommon for wind turbines to face opposition and skepticism from local communities. This section explores the various factors influencing public perception and acceptance of wind turbines, as well as strategies for addressing concerns and promoting positive attitudes towards this renewable energy source.

Understanding Public Concerns

When it comes to wind turbines, the concerns voiced by the public often revolve around three main areas: visual impact, noise pollution, and health effects. Let's delve deeper into each of these concerns:

Visual Impact: Wind turbines, with their towering structures and rotating blades, are often seen as eyesores that disrupt the natural beauty of landscapes. Some people argue that they degrade scenic views and negatively impact tourism. Additionally, concerns over the effect of wind turbines on property values are also commonly raised.

Noise Pollution: The noise generated by wind turbines has been a cause for concern among nearby residents. While modern wind turbines are designed to minimize noise, the low-frequency sounds they produce can still be audible to some individuals, especially in quiet rural areas. People worry that this noise can lead to sleep disturbances, stress, and other health issues.

Health Effects: Another concern raised by critics is the potential health effects associated with wind turbines. Some individuals claim that the low-frequency noise emitted by wind turbines can cause headaches, dizziness, and other adverse health effects. However, scientific studies have generally not found any direct causal link between wind turbine operation and negative health outcomes.

Addressing Concerns and Building Acceptance

To address public concerns and foster acceptance of wind turbines, it is essential to engage with affected communities, provide accurate information, and implement effective mitigation strategies. Here are some approaches to consider:

Community Engagement: Engaging with local communities early in the planning stages of wind turbine projects is crucial. By involving residents in the decision-making process, their concerns can be heard and addressed. Public forums, open houses, and community consultations provide opportunities for dialogue and transparency.

Visual Integration and Design: Minimizing the visual impact of wind turbines can help alleviate concerns. Striving for aesthetically pleasing designs and locating turbines strategically where they are less obtrusive can go a long way in garnering acceptance. Utilizing smaller turbines or exploring alternatives like vertical axis wind turbines can also reduce the visual impact.

Noise Mitigation: Implementing measures to mitigate noise pollution is essential. Modern wind turbine designs focus on reducing noise levels, but additional steps can be taken to further minimize any disturbances. This may include implementing setback distances, utilizing sound barriers, or incorporating advanced noise-reduction technologies.

Education and Communication: Effective communication is key to building public acceptance. Providing accurate information about wind energy, addressing misconceptions, and highlighting the benefits of clean, renewable energy sources can help dispel fears and foster positive attitudes. Educational outreach programs, workshops, and online resources can be valuable tools in this regard.

NOISE AND HEALTH CONCERNS

Community Benefits: Demonstrating the economic and environmental benefits that wind turbines can bring to local communities is essential. Job creation, revenue generation, and local investment opportunities associated with wind energy projects can help build support. Additionally, exploring community ownership models, where residents have a stake in the project, can further enhance acceptance.

Research and Monitoring: Continued research on the impacts of wind turbines, including noise levels, wildlife interactions, and health effects, is essential to address concerns in an evidence-based manner. Ongoing monitoring and evaluation of existing wind projects can help identify potential issues and inform best practices.

Example: The Wind Energy Divide

To illustrate the complexities of public perception and acceptance, let's consider the case of a proposed wind farm in a coastal community. The project promises to bring renewable energy, reduce carbon emissions, and boost the local economy. Despite these potential benefits, some residents are vehemently opposed to the wind farm due to concerns over visual impact, noise, and potential health effects.

In response, the project developers engage with the community through public consultations and conduct thorough environmental assessments. They address concerns by redesigning the layout of the wind farm to minimize visual impact, implementing noise reduction measures, and providing detailed information on the studies that debunk claims of health risks.

Additionally, the developers involve community members in the decision-making process by exploring ownership options that would allow local residents to benefit financially from the wind farm. They also work closely with environmental organizations to ensure that wildlife habitats are protected and any potential environmental impacts are mitigated.

Over time, through open dialogue, education, and careful consideration of community concerns, the perception of the wind farm begins to shift. The once-opposed residents start to appreciate the project's contribution to renewable energy, local job creation, and economic growth. They understand that wind turbines are a necessary step towards a sustainable future, and their initial concerns are gradually replaced with acceptance and support.

By actively addressing public concerns, engaging with communities, and providing accurate information, it is possible to bridge the gap between opposition and acceptance, leading to the successful implementation of wind turbine projects.

Further Reading and Resources

For further exploration of the topic of public perception and acceptance of wind turbines, the following resources are recommended:

- *Wind Energy Explained* by James F. Manwell, Jon G. McGowan, and Anthony L. Rogers.

- *Wind Energy Handbook* by Tony Burton, Nick Jenkins, David Sharpe, and Ervin Bossanyi.

- *Engaging Communities in Wind Energy Projects: The Respect Model and Beyond* by Patrick Devine-Wright.

- The U.S. Department of Energy's Wind Energy Technologies Office website: `https://www.energy.gov/eere/wind/wind-energy-technologies-office`

- The American Wind Energy Association website: `https://www.awea.org`

Remember, the journey towards widespread acceptance of wind turbines requires ongoing conversations, open-mindedness, and a commitment to sustainability. By addressing concerns, fostering dialogue, and emphasizing the benefits of renewable energy, we can pave the way for a greener and more prosperous future.

Maintenance and Longevity

Structural Issues and Component Failures

When it comes to wind turbines, structural issues and component failures are two major concerns that need to be addressed. These problems can lead to reduced efficiency, increased costs, and in some cases, even catastrophic failures. In this section, we will explore the common structural issues and component failures that can occur in wind turbines, and discuss the impact they have on the overall performance and safety of these renewable energy systems.

Structural Issues

Structural issues in wind turbines primarily revolve around the rotor blades, tower, and foundation. Let's take a closer look at each of these components.

MAINTENANCE AND LONGEVITY

Rotor Blades The rotor blades are arguably the most critical component of a wind turbine, as they capture the energy from the wind and convert it into rotational motion. However, these blades are subjected to significant stresses and strains during operation, which can lead to fatigue, material degradation, and ultimately, failure. Some common structural issues with rotor blades include:

- **Leading Edge Erosion:** Wind turbines are often installed in areas with harsh weather conditions, such as coastal regions. The leading edge of the rotor blades can be damaged by rain, hail, or airborne particles, leading to reduced aerodynamic performance and structural integrity.

- **Delamination:** Rotor blades are typically made of composite materials, such as fiberglass or carbon fiber reinforced polymers. Delamination, which refers to the separation of different layers within the blade structure, can occur due to manufacturing defects, impact damage, or prolonged exposure to environmental factors such as moisture and UV radiation.

- **Fatigue Cracking:** Wind turbines constantly experience cyclic loading due to variations in wind speed and direction. Over time, this cyclic loading can lead to the initiation and propagation of fatigue cracks in the rotor blades, compromising their strength and lifespan.

To mitigate these structural issues, continuous monitoring and maintenance of rotor blades are essential. Regular inspections, repair of minor damages, and timely replacement of severely damaged or worn-out blades can help avoid catastrophic failures and optimize the performance of wind turbines.

Tower The tower provides support and stability to the entire wind turbine structure. However, it is also prone to several structural issues, including:

- **Tower Oscillations:** Wind-induced vibrations can cause the tower to oscillate, leading to fatigue and potential failure over time. The natural frequency of the tower must be carefully designed to avoid resonance with the wind excitation frequency.

- **Corrosion:** Towers are exposed to environmental conditions, such as moisture and saltwater in coastal areas, which can accelerate corrosion of the structural components. Corrosion weakens the tower's integrity and reduces its load-bearing capacity.

- **Foundation Settlement:** Wind turbines require a solid foundation to withstand the loads imposed by the rotor and tower. Improper design or construction of the foundation can result in settlement over time, leading to misalignment, increased stress on the tower, and potential structural failures.

Regular inspections, maintenance, and corrosion protection measures can help mitigate these structural issues. Additionally, advancements in tower design, such as tubular steel or concrete hybrid towers, can improve their durability and longevity.

Component Failures

Apart from the structural issues mentioned above, various components within a wind turbine can also fail, impacting the overall performance and reliability of the system. Let's explore some common component failures in wind turbines.

Gearbox The gearbox is a critical component that connects the low-speed shaft from the rotor to the high-speed shaft of the generator, increasing the rotational speed. However, it is also one of the most failure-prone components in wind turbines. Some common gearbox failures include:

- **Gear Tooth Breakage:** High loads and rapid changes in torque can lead to fatigue failure of gear teeth, causing them to break and resulting in a complete loss of power generation.

- **Bearing Failure:** Excessive loads, inadequate lubrication, or manufacturing defects can cause bearing failures in the gearbox. Bearing failures lead to increased friction, heat generation, and vibration, ultimately resulting in a breakdown of the entire gearbox system.

To mitigate gearbox failures, manufacturers have been exploring alternative drivetrain designs, such as direct-drive systems, which eliminate the need for a gearbox altogether. These designs are more reliable and require less maintenance.

Generator The generator converts the rotational energy of the wind turbine into electrical energy. However, it can also experience failures, including:

- **Electrical Faults:** Generator failures can occur due to electrical faults, such as short circuits or insulation breakdown. These faults can result in interruptions in power generation or even cause extensive damage to the generator.

- **Cooling System Failure:** Generators produce a significant amount of heat during operation, requiring an efficient cooling system to maintain their temperature within acceptable limits. Cooling system failures can lead to overheating, causing the generator to shut down or sustain damage.

Regular maintenance and monitoring of the generator, along with effective cooling systems and electrical protection measures, can minimize the risk of generator failures.

Yaw System   The yaw system controls the orientation of the wind turbine rotor, ensuring it faces the wind direction for optimal energy capture. Yaw system failures can have serious consequences, including:

- **Yaw Drive Failure:** The yaw drive rotates the nacelle (housing the rotor and other critical components) to align it with the wind direction. Failures in the yaw drive can result in misalignment, reduced power output, and increased stress on other components.

- **Yaw Braking System Failure:** The yaw braking system is responsible for locking the wind turbine in position during maintenance or in case of emergencies, such as high wind speeds. If the braking system fails, the wind turbine may lose control and potentially become a safety hazard.

Proper maintenance, regular inspection, and repair of the yaw system can help prevent these failures and ensure the safe and efficient operation of wind turbines.

Conclusion

Structural issues and component failures are significant challenges in the operation and maintenance of wind turbines. Understanding and addressing these issues are crucial for optimizing the performance, safety, and longevity of these renewable energy systems. By implementing effective maintenance strategies, incorporating advanced materials and designs, and adopting innovative solutions, we can overcome these challenges and harness the full potential of wind energy. Remember, the success of wind turbines not only lies in capturing the power of the wind but also in their structural resilience and robust components.

Maintenance Challenges in Remote Locations

Maintaining wind turbines in remote locations poses a unique set of challenges. In this section, we will explore the various obstacles faced by maintenance teams operating in these areas and discuss the strategies employed to overcome them.

Access and Logistics

Remote locations often lack proper infrastructure, making it difficult to reach wind turbine sites. Rough terrains, inaccessible roads, and limited transportation options can impede timely maintenance. In such cases, maintenance crews rely on specialized equipment, like all-terrain vehicles (ATVs) and helicopters, to access these sites.

Helicopter transport plays a crucial role in servicing turbines located in remote and offshore areas. It allows technicians to quickly reach the site, carry out repairs or inspections, and return without the need for time-consuming land travel. Helicopters also enable the transportation of heavy tools and replacement parts, minimizing downtime caused by logistics challenges.

To optimize maintenance efforts in remote locations, pre-planning and efficient coordination are key. Maintenance schedules are carefully organized, considering multiple turbine visits in one trip and ensuring that all required tools and spare parts are available. This approach minimizes the number of visits required, reducing costs and increasing overall efficiency.

Harsh Environmental Conditions

Remote locations are often characterized by extreme weather conditions, which pose additional maintenance challenges. For instance, turbines situated in offshore or coastal areas are subjected to salty air, high wind speeds, and corrosive conditions that can degrade critical components.

To mitigate these challenges, wind turbine manufacturers employ specialized materials and coatings that enhance corrosion resistance. Regular cleaning and inspection of turbine blades are also necessary to remove dirt, salt deposits, or ice buildup, which can affect aerodynamic performance. Remote monitoring systems equipped with weather sensors are employed to gather real-time data and detect any anomalies that may require immediate attention.

Additionally, maintenance teams in remote locations must be equipped with proper safety gear and training to operate in harsh environments. This includes protective clothing, harnesses, and training on working at heights and in adverse weather conditions.

Limited Resources

Remote locations often have limited access to resources such as electricity, water, and skilled technicians. This scarcity adds complexity to maintenance operations, as technicians must rely on alternative energy solutions and carefully manage available resources.

To address this challenge, wind turbine sites in remote areas are often equipped with off-grid power systems, such as solar panels or diesel generators, to supply electricity for maintenance activities. Water usage is optimized through the implementation of water recycling systems and water-efficient cleaning techniques.

Skilled technicians are a valuable resource, and their presence in remote locations may be limited. To overcome this, local community involvement and training programs can be established to develop a pool of skilled technicians from the nearby area. This approach not only ensures a readily available workforce but also boosts local employment opportunities and fosters a sense of ownership and responsibility within the community.

Remote Monitoring and Predictive Maintenance

Remote monitoring systems play a vital role in maintenance operations in remote locations. These systems enable real-time monitoring of turbine performance, allowing technicians to identify potential issues before they become critical failures.

Sensor-equipped components collect data on various parameters such as temperature, vibration, and energy production. These data are transmitted to a central control system where they are analyzed to detect anomalies or deviations from normal behavior. Technicians can then prioritize maintenance activities based on criticality and plan interventions accordingly.

Predictive maintenance strategies, combined with remote monitoring systems, help optimize the scheduling of maintenance activities. By predicting component lifespans and identifying potential equipment failures in advance, maintenance teams can plan and execute repairs or replacements efficiently, minimizing downtime and reducing overall maintenance costs.

Unconventional Solution: Drones for Inspections

An unconventional yet valuable solution for maintenance in remote locations is the use of drones for inspections. Drones equipped with high-resolution cameras and sensors can provide detailed and rapid visual inspections of turbine components, including blades, nacelles, and towers.

By conducting routine inspections using drones, potential issues such as blade erosion, cracks, or loose connections can be identified early. The collected data can be analyzed to assess the condition of the turbine and plan targeted maintenance activities, reducing the need for intrusive manual inspections and optimizing maintenance efforts.

Drones also provide an efficient means of inspecting tall structures in challenging terrains, such as mountains or dense forests, where traditional access methods are impractical or unsafe. They enable maintenance teams to perform inspections quickly, accurately, and cost-effectively.

In conclusion, maintenance in remote locations presents unique challenges that require innovative solutions. Access and logistics, harsh environmental conditions, limited resources, and the use of remote monitoring systems and predictive maintenance strategies are essential considerations. Furthermore, the utilization of drones for inspections offers an unconventional yet efficient approach to maintaining wind turbines in remote areas. By addressing these challenges head-on, maintenance teams can ensure the reliability and longevity of wind turbines, contributing to the success of renewable energy endeavors.

Decommissioning and Waste Management

Decommissioning wind turbines is an important process that involves the dismantling and removal of the turbines once they have reached the end of their operational life. Additionally, waste management plays a critical role in ensuring that any waste generated during the decommissioning process is properly handled and disposed of in an environmentally responsible manner.

Decommissioning Process

The decommissioning process for wind turbines involves several key steps to ensure safe and efficient removal. These steps include:

1. **Planning and Preparation**: Before the decommissioning process begins, a detailed plan is developed to outline the necessary steps and resources required. This includes obtaining any necessary permits or approvals and ensuring that safety measures are in place.

2. **Dismantling of Components**: The first step in decommissioning is the dismantling of the turbine components. This includes removing the rotor, nacelle, and tower. Specialized equipment and skilled technicians are typically employed for this task to ensure safe and efficient removal.

3. **Transportation and Disposal:** Once the turbine components have been dismantled, they need to be carefully transported to an appropriate disposal facility. This may involve transporting them to a recycling center or a designated landfill site. Transportation methods need to be planned to minimize any potential risks to the environment or public safety.

4. **Site Restoration:** After the turbine components have been removed, the site needs to be restored to its original condition as much as possible. This may involve regrading the land, removing any temporary infrastructure, and replanting vegetation to restore the natural habitat.

5. **Environmental Assessments:** Throughout the decommissioning process, environmental assessments should be conducted to ensure that any potential environmental impacts are identified and mitigated. This includes assessing the potential impacts on soil, water, air quality, and local biodiversity.

6. **Monitoring and Reporting:** Following the decommissioning process, ongoing monitoring and reporting should be conducted to ensure that any potential issues or impacts are identified and addressed promptly. This includes monitoring water quality, air quality, and wildlife populations in the area surrounding the decommissioned site.

Waste Management

Waste management is an important aspect of decommissioning wind turbines, as it ensures that any waste generated during the process is handled properly and does not pose a threat to the environment or human health. The waste generated during decommissioning can be categorized into three main types:

1. **Metal Waste:** Wind turbines contain a significant amount of metal components, such as steel and aluminum. These metals can be recycled and reused, reducing the need for new raw materials. Proper sorting and recycling of metal waste is crucial to minimize environmental impacts and conserve resources.

2. **Non-Metal Waste:** Non-metal waste includes materials such as plastic, fiberglass, and composites. While these materials may not be suitable for recycling in their current form, efforts should be made to explore alternative uses or environmentally friendly disposal methods. Research and development in the field of composite recycling are ongoing to find effective solutions for the management of non-metal waste.

3. **Hazardous Waste:** Certain components of wind turbines, such as lubricants, hydraulic fluids, and electrical equipment, may contain hazardous substances. Proper handling and disposal of hazardous waste are essential to prevent pollution and minimize any potential risks to human health and the environment. Strict adherence to regulations and guidelines for hazardous waste management is necessary.

It is important to note that effective waste management strategies for decommissioned wind turbines require collaboration between stakeholders, including turbine manufacturers, contractors, waste management facilities, and regulatory agencies. Furthermore, the development of innovative technologies for recycling and reusing turbine components can significantly reduce the environmental impact of decommissioning.

Case Study: Recycling and Reusing Wind Turbine Components

As the number of decommissioned wind turbines continues to grow, finding sustainable solutions for recycling and reusing their components is becoming increasingly important. One case study that highlights the potential for recycling is the Repowering Project in Denmark.

The Repowering Project aimed to replace older, less efficient wind turbines with new and more advanced models. To minimize waste and maximize resource efficiency, the project embraced a circular economy approach. This involved the careful dismantling of the decommissioned turbines, with a particular focus on recycling and reusing components.

During the dismantling process, valuable materials such as steel, copper, and aluminum were sorted and sent for recycling. The remaining components, including blades made of composite materials, were carefully processed to explore alternative uses. For example, the blades were crushed and used as raw material in the production of cement, thereby reducing the need for virgin materials. This innovative approach not only reduced waste but also contributed to a more sustainable construction industry.

The Repowering Project serves as an excellent example of how careful planning, resource efficiency, and collaboration among stakeholders can lead to successful decommissioning with minimal environmental impact. It showcases the importance of exploring innovative solutions and developing a circular economy mindset to address the waste management challenges associated with wind turbine decommissioning.

Caveats and Challenges

While efforts are being made to develop effective decommissioning and waste management strategies for wind turbines, several challenges remain. These challenges include:

- **Lack of Standardization:** Currently, there is no standardized approach to wind turbine decommissioning and waste management. This lack of standardization can make it difficult to ensure consistent practices across different projects and jurisdictions.

- **Technological Advances:** Wind turbine technology is rapidly evolving, which can pose challenges for decommissioning. As turbines become larger and incorporate new materials, it is essential to stay updated on the latest techniques and technologies for dismantling and recycling these advanced systems.

- **Economic Viability:** The economic viability of decommissioning and waste management practices is a significant consideration. Balancing the costs of decommissioning, transportation, and recycling with the value of the recovered materials can be complex and may require innovative financial models.

- **Regulatory Frameworks:** Comprehensive regulatory frameworks addressing decommissioning and waste management for wind turbines are still in the early stages of development. These frameworks need to be robust, flexible, and adaptable to ensure the safe and environmentally responsible decommissioning of wind turbines.

- **Public Perception:** The public perception of wind turbine decommissioning and waste management can impact the overall acceptance and support for renewable energy projects. It is crucial to engage with local communities and stakeholders to address any concerns and provide transparent information regarding the decommissioning process and waste management practices.

Despite these challenges, ongoing research, collaboration, and innovation are expected to lead to improved decommissioning practices and more sustainable waste management solutions for wind turbines.

Conclusion

Decommissioning wind turbines and effectively managing the associated waste are critical steps in the lifecycle of renewable energy infrastructure. The decommissioning process, including careful planning, dismantling, transportation, and site restoration, ensures the safe removal of wind turbines. Waste management, particularly recycling and proper disposal of metal, non-metal, and hazardous waste, minimizes environmental impacts and maximizes resource efficiency.

While challenges exist, such as the lack of standardization and evolving technology, ongoing efforts are being made to address these issues and develop sustainable decommissioning and waste management strategies. Public engagement and collaboration among stakeholders will be crucial to ensure the successful decommissioning of wind turbines and the transition to a more circular and sustainable renewable energy industry.

By embracing the principles of the circular economy, exploring innovative recycling techniques, and developing comprehensive regulatory frameworks, the decommissioning and waste management of wind turbines can be conducted in an environmentally responsible and socially beneficial manner. Through these efforts, the renewable energy industry can continue to contribute to global efforts in combating climate change and transitioning towards a more sustainable future.

Case Studies and Controversies

Cape Wind Project

Background and Opposition

Historical Development of Wind Turbines

To understand the background and opposition to wind turbines, we need to delve into their historical development. Wind turbines have been used for centuries, dating back to ancient civilizations that used wind power to grind grain and pump water. The modern wind turbine, however, has a more recent origin.

The first electricity-generating wind turbine was developed by Charles F. Brush in the late 19th century. This early design consisted of a large rotor with multiple blades connected to an electrical generator. While it produced electricity, it was not efficient enough to compete with fossil fuel-based power generation.

It wasn't until the 1970s, during the oil crisis, that interest in wind energy resurged. The United States initiated a research program to develop wind turbines for utility-scale power generation. The Danish also became pioneers in wind energy, leading to the development of the first large-scale wind farms in the early 1980s.

Since then, wind turbine technology has improved significantly, with larger and more efficient turbines capable of generating substantial amounts of clean electricity. These advancements have made wind energy a viable option for meeting the growing global demand for renewable energy.

Importance of Wind Turbines in Renewable Energy

Wind turbines play a vital role in the transition to a sustainable and low-carbon future. Clean and renewable energy sources are essential for reducing greenhouse gas emissions and mitigating climate change. Wind energy, along with solar and hydroelectric power, is one of the most readily available and scalable renewable energy options.

Wind power offers several advantages over conventional fossil fuel-based electricity generation. It is a clean energy source that produces no direct emissions or air pollutants. Additionally, it is a domestic energy source that reduces dependence on imported fossil fuels, enhancing energy security.

Wind turbines contribute to global efforts to combat climate change by displacing carbon dioxide emissions from traditional power plants. They help reduce the environmental footprint associated with electricity generation and provide a sustainable alternative to non-renewable energy sources.

Opposition to Wind Turbines

Despite their numerous advantages, wind turbines have faced opposition from various stakeholders. Understanding the concerns and addressing them is crucial for the widespread acceptance and deployment of wind energy.

One of the primary oppositions to wind turbines is their visual impact on landscapes. Some argue that wind turbines disrupt natural beauty, particularly in scenic or culturally significant areas. Critics claim that these structures harm tourism, which often relies on unspoiled landscapes as attractions.

Noise pollution is another commonly raised concern. While modern wind turbines are designed to minimize noise, some individuals living in close proximity to wind farms may experience annoyance due to the low-frequency noise generated by the rotating blades. Although the health effects of such noise remain a topic of debate, it is important to address community concerns and ensure proper sound mitigation measures are implemented.

Bird and bat fatalities have also been a point of contention in the opposition to wind turbines. The rotating blades can pose a threat to these species, particularly in areas with high bird or bat populations. Steps must be taken to mitigate these risks, such as carefully selecting turbine locations and implementing avian deterrent measures.

Lastly, there may be concerns about the potential impact on property values near wind farms. Some argue that the presence of wind turbines may decrease property values, making it harder for homeowners to sell their properties or obtain fair compensation.

While these concerns should not be dismissed, it is important to consider the broader context. The environmental benefits and economic opportunities associated with wind energy are significant. By addressing the concerns through responsible planning, technological advancements, and stakeholder engagement, it is possible to strike a balance and continue expanding wind energy as a valuable renewable resource.

Further Reading

For a comprehensive analysis of the opposition to wind turbines, we recommend the following resources:

- "Wind Energy and the Impact on Wildlife: Addressing Wildlife Concerns and Siting Wind Turbines" by Martin E. Wieland and J. Maarten de Groot.

- "The Impact of Wind Energy Development on Landscapes: A Review" by Danaë M. A. Rozendaal and Frank van Langevelde.

- "Wind Farm Noise: Annoyance, Sleep Disturbance, and Environmental Impacts" by Geoffrey G. Leventhall, B. Alana-Selle McCombie, and Richard R. Hooper.

- "Wind Turbines and Birds: A Guidance Document for Environmental Assessment" by The American Wind Energy Association and The Wildlife Society.

- "Wind Farm Development and Property Values" by Ben Hoen, Ryan Wiser, and Peter Cappers.

These references provide a balanced perspective on the opposition to wind turbines, incorporating scientific research, industry best practices, and real-world case studies. Through a deeper understanding of the issues, we can continue the dialogue and work towards a more sustainable and inclusive energy future.

Legal and Regulatory Challenges

Dealing with legal and regulatory challenges is an inevitable aspect of any major infrastructure project, and wind turbines are no exception. From obtaining permits to addressing concerns of local communities, there are various legal and regulatory hurdles that need to be navigated in order to successfully develop and operate wind farms. In this section, we will explore the key legal and regulatory challenges that wind turbine projects often face and discuss potential solutions to overcome them.

Permitting and Environmental Impact Assessments

Before the construction of a wind farm can begin, developers must obtain the necessary permits from regulatory authorities. This process often involves conducting an environmental impact assessment (EIA) to evaluate the potential effects of the project on the surrounding ecosystem. EIAs are crucial for ensuring that wind farms comply with environmental regulations and are built in a sustainable manner.

However, the EIA process can be time-consuming and costly, often resulting in project delays. In some cases, environmental organizations and local communities may raise concerns about the potential negative impacts of wind turbines on wildlife, habitats, and landscapes. These concerns can lead to legal challenges and further elongate the permitting process.

To address these challenges, it is important for wind turbine developers to engage in early and transparent communication with local stakeholders, including environmental groups and communities. By addressing concerns and providing accurate and up-to-date information about the environmental impacts of wind farms, developers can build trust and minimize opposition.

Land Rights and Zoning Regulations

Another significant challenge in wind turbine projects is acquiring appropriate land rights and navigating zoning regulations. Wind farms require vast areas of land, and securing the necessary land rights can be complex, particularly when dealing with multiple landowners.

In some instances, landowners may be reluctant to lease their land due to concerns about potential negative impacts such as noise, visual pollution, or reduced property values. Moreover, conflicts may arise when constructing wind farms near residential areas or within areas designated for other purposes, such as protected natural areas or historical sites.

To overcome these challenges, developers can consider implementing community engagement programs and offer fair compensation to landowners. By involving local communities in the decision-making process and addressing their concerns, developers can foster a positive relationship and establish a constructive dialogue.

Additionally, adhering to zoning regulations and working closely with local planning authorities is crucial. Zoning regulations determine where wind farms can be located and what restrictions may apply. Understanding and complying with these regulations can help streamline the permitting process and mitigate potential legal obstacles.

Electricity Grid Connection and Market Regulations

Wind farms need to be connected to the electricity grid to distribute the generated power. However, connecting wind farms to the grid can present challenges, particularly in remote areas or in regions with insufficient grid infrastructure. Grid connection may require additional investments in transmission lines and substations, adding complexity and cost to the project.

Moreover, market regulations and policies can impact the economic viability of wind farms. Some regions have complex electricity market structures that might not fully recognize or reward the value of renewable energy. This can create uncertainty regarding the long-term profitability of wind turbine projects.

To address these challenges, policymakers and grid operators should prioritize grid expansion and modernization to accommodate the integration of renewable energy sources. Furthermore, supportive policies such as feed-in tariffs or renewable portfolio standards can provide stable and predictable revenue streams, encouraging investment in wind farms.

Legal Disputes and Litigation

Unfortunately, legal disputes and litigation can arise during various stages of wind turbine projects. These disputes can include contract disagreements, land rights disputes, or concerns raised by local communities.

To minimize the risk of legal challenges, developers should ensure that they have robust legal contracts in place, clearly defining the rights and responsibilities of all parties involved. Engaging experienced legal counsel during project development can help anticipate and mitigate potential legal risks.

Additionally, developers should proactively work with local communities, addressing concerns, and striving for mutually beneficial solutions. By fostering open communication and transparency, disputes can often be resolved through negotiation and collaboration rather than resorting to litigation.

Conclusion

Navigating the legal and regulatory landscape is a vital aspect of any wind turbine project. From permitting and environmental impact assessments to addressing land rights issues and grid connection challenges, there are numerous legal and regulatory challenges that must be addressed.

Developers can overcome these challenges by prioritizing early and transparent communication with stakeholders, implementing community engagement programs, and offering fair compensation to landowners. Additionally, collaborating with policymakers and grid operators to improve grid infrastructure and establish supportive policies can enhance the economic viability of wind farms.

By understanding and addressing the legal and regulatory challenges associated with wind turbines, developers can navigate the complex landscape and ensure the successful development and operation of sustainable and environmentally friendly wind farms.

Lessons Learned from Cape Wind

The Cape Wind Project, proposed in 2001, was a plan to build a wind farm off the coast of Cape Cod, Massachusetts. It aimed to harness the consistent sea breezes of Nantucket Sound to generate clean, renewable energy. However, the project faced numerous challenges and controversies, leading to important lessons that can be applied to future wind energy initiatives.

5.1.3.1 Stakeholder Engagement and Community Support

One of the key lessons learned from the Cape Wind Project is the importance of early and effective stakeholder engagement. The project faced considerable opposition from local residents, including prominent individuals such as the Kennedy family, who raised concerns about the impact on scenic views and property values.

To overcome these challenges, it is crucial for wind energy developers to engage with the local community from the start. Open and transparent communication, addressing concerns, and actively involving the community in the planning process can help gain their support.

Furthermore, educating the public about the benefits of wind energy and dispelling any misconceptions can help build trust and generate positive public sentiment. This can be achieved through public forums, informational campaigns, and participation in local events.

5.1.3.2 Legal and Regulatory Hurdles

The Cape Wind Project struggled with multiple legal and regulatory obstacles. The permitting process was protracted, with opponents filing lawsuits to block the project on grounds such as environmental impact, navigational safety, and failure to comply with established regulations.

To avoid similar challenges, wind energy projects must have a clear understanding of the applicable legal and regulatory frameworks. Developing relationships with regulatory agencies, ensuring compliance with all relevant regulations, and actively addressing environmental concerns can minimize the risk of lawsuits and other legal roadblocks.

It is also important for wind energy developers to understand the potential impacts of their projects and undertake thorough environmental assessments. This includes conducting studies on wildlife, marine ecosystems, and visual landscapes, while actively involving local experts and stakeholders. By proactively addressing potential impacts, developers can present a strong case for their projects and demonstrate their commitment to environmental stewardship.

5.1.3.3 Lessons in Public Perception and Acceptance

The Cape Wind Project highlighted the significance of public perception and acceptance. It underscored the need for wind energy developers to proactively address concerns related to noise, visual pollution, and potential health effects, and to communicate the benefits of wind energy effectively.

Investing in studies that assess noise levels, especially low-frequency noise, and sharing the results with the public can help alleviate concerns. Additionally, implementing mitigation measures, such as setback requirements, can help minimize the impact of noise on nearby residents.

Visual pollution is another factor that can influence public perception. Wind energy developers should consider the aesthetic aspects of their projects, exploring design options that integrate with the surrounding environment. In the case of Cape Wind, considerable emphasis on the visual impact influenced public opinion and resistance.

Lastly, addressing potential health concerns related to wind turbines is essential. While no substantive evidence suggests that wind turbine noise causes adverse health effects, perceptions and concerns persist. Developers should actively communicate the scientific consensus on this issue while acknowledging and addressing individual experiences.

5.1.3.4 Economic Considerations and Project Viability

The Cape Wind Project faced criticism due to its high cost, which raised questions about the economic viability of offshore wind energy. The project required substantial investment in infrastructure, transmission lines, and other associated costs, contributing to a higher price of the generated electricity.

To enhance the economic viability of wind energy projects, it is crucial to leverage technological advancements and economies of scale. The development of larger and more efficient turbines, as well as exploring alternative designs like floating and vertical-axis wind turbines, can help reduce costs and improve energy generation.

Moreover, wind energy developers should consider potential revenue streams beyond electricity sales. For example, collaborations with local industries, such as fishing or tourism, can create additional economic benefits, contributing to community support.

5.1.3.5 Collaboration and Learning from International Projects

Considering the global growth of wind energy, it is crucial to learn from international experiences. The Cape Wind Project provided valuable lessons for future offshore wind farm initiatives, not only within the United States but globally as well.

Collaborating with established offshore wind sectors, such as those in Europe, can facilitate knowledge sharing and best practices exchange. Understanding successful international projects can provide insights into effective planning, stakeholder engagement, and environmental impact mitigation strategies.

By building on existing expertise and leveraging lessons learned from international projects, wind energy developers can streamline the development process and increase the chances of project success.

Conclusion

The Cape Wind Project serves as a significant case study in the challenges and lessons associated with wind energy development. It highlighted the importance of stakeholder engagement, adherence to legal and regulatory frameworks, addressing public perception, and ensuring economic viability.

By applying these lessons, future wind energy projects can navigate the complexities and controversies associated with renewable energy development. Ultimately, these lessons can contribute to the effective deployment of wind energy as a sustainable and vital component of our clean energy future.

Offshore Wind Farms

Advantages and Disadvantages of Offshore Wind

Offshore wind farms have gained significant attention in recent years as a promising source of renewable energy. This section will explore the advantages and disadvantages associated with offshore wind, providing a comprehensive overview of the topic.

Advantages of Offshore Wind

1. **Vast and consistent wind resources:** Offshore areas generally experience stronger and more consistent winds compared to onshore sites. The absence of

obstructions, such as buildings and vegetation, allows for a higher wind speed and a greater potential for electricity generation.

2. **Higher energy yields:** The stronger and more consistent winds offshore result in higher energy yields from wind turbines. Offshore wind turbines have larger rotor diameters, enabling them to capture more wind energy and convert it into electricity. As a result, offshore wind farms have higher capacity factors and can generate more electricity over time compared to onshore wind farms.

3. **Reduced visual impact:** One of the main advantages of offshore wind farms is their reduced visual impact. By being located far from the coast, they are less visible from shorelines and residential areas. This mitigates the aesthetic concerns often associated with onshore wind turbines, making offshore wind more socially acceptable in some communities.

4. **Access to stronger winds near coastlines:** Offshore wind farms can be situated close to densely populated coastal areas, where electricity demand is typically high. By tapping into the strong winds near these areas, offshore wind can provide clean and reliable energy to meet the needs of coastal communities without the need for long-distance transmission.

5. **Potential for larger-scale installation:** Offshore wind farms have the potential for larger-scale installation compared to onshore wind farms. With vast expanses of open ocean available, it is possible to deploy a significant number of wind turbines, creating megawatt-scale projects that have a greater overall capacity to generate electricity.

6. **Potential for technological advancement:** Offshore wind technology is still relatively new and evolving. As more projects are developed and lessons are learned, there is significant potential for technological advancement in areas such as turbine design, installation techniques, and maintenance strategies. This continuous innovation can further improve the performance and cost-effectiveness of offshore wind energy.

Disadvantages of Offshore Wind

1. **High capital and maintenance costs:** The installation and maintenance of offshore wind farms involve significant capital expenditures. Building and operating turbines in harsh marine environments require specialized engineering, equipment, and skilled labor, all of which contribute to higher costs compared to onshore wind projects. Additionally, the cost of maintaining turbines in deep waters can be more challenging and expensive.

2. **Technical challenges and risks:** Offshore wind farms face unique technical challenges and risks. Harsh weather conditions, such as storms and high waves,

can adversely affect the performance and reliability of wind turbines. The corrosive nature of seawater also poses challenges for the durability and longevity of offshore wind infrastructure. Mitigating these technical challenges requires robust engineering solutions and ongoing maintenance.

3. **Environmental concerns**: Offshore wind farms can have environmental impacts, although these are generally less compared to other forms of energy generation. Construction activities, such as pile driving for turbine foundations, can create underwater noise that may disturb marine life. The presence of turbines can also lead to habitat displacement and potential risks to marine species, particularly during their migratory routes. However, careful project planning and environmental impact assessments can help minimize these effects.

4. **Grid connection challenges**: Connecting offshore wind farms to the onshore grid requires the installation of submarine cables, which can be technically challenging, expensive, and subject to potential failures. The large distance between offshore wind farms and onshore substations can result in transmission losses and grid instability. Furthermore, when multiple offshore wind farms are connected to a single grid, the intermittent nature of wind can pose challenges in balancing electricity supply and demand.

5. **Potential conflicts with other marine activities**: Offshore wind farms occupy significant marine space, which can create conflicts with other marine activities such as fishing, shipping, and recreational boating. The proximity of wind farms to shipping channels and fishing grounds can raise concerns related to navigational safety and the disruption of established industries. Effective spatial planning and stakeholder engagement are crucial to finding a balance between offshore wind energy and other marine uses.

6. **Limited experience and knowledge base**: Compared to onshore wind, the offshore wind industry is still relatively young, resulting in a limited experience and knowledge base. This can lead to uncertainties and complexities in project development, from site selection to turbine installation and maintenance. Expanding the offshore wind sector requires continued research, collaboration, and knowledge sharing to address these challenges.

In conclusion, offshore wind presents several advantages, such as access to stronger and more consistent winds, higher energy yields, and reduced visual impact. However, it also has inherent disadvantages, including high costs, technical challenges, environmental concerns, grid connection issues, potential conflicts with other marine activities, and a limited experience base. Despite these challenges, the rapidly evolving technology and increasing interest in offshore wind signify its potential to contribute significantly to the global renewable energy transition.

Challenges and Environmental Concerns

Wind turbines have gained popularity as a renewable energy source due to their ability to generate electricity without producing greenhouse gas emissions. However, like any form of energy production, wind turbines come with their own set of challenges and environmental concerns. In this section, we will explore some of these challenges and discuss their potential impacts on the environment.

Intermittency and Energy Storage

One of the main challenges with wind energy is its intermittent nature. Wind speed is not constant, and therefore, the amount of electricity generated by wind turbines fluctuates. This intermittency poses problems for grid integration and stability, as the supply of wind energy may not always match the demand.

To overcome this challenge, energy storage technologies play a crucial role. Energy storage systems, such as batteries, can store excess electricity generated during periods of high wind activity and release it during periods of low wind or increased demand. These systems help bridge the gap between supply and demand, ensuring a steady and reliable energy supply.

However, energy storage technologies are still relatively expensive and have limited capacity. The development of cost-effective and efficient energy storage solutions remains a challenge for the wind energy industry.

Environmental Impact on Wildlife

Although wind energy is considered environmentally friendly, it is not entirely without negative impacts on wildlife. One of the major concerns is the threat wind turbines pose to birds and bats.

Birds, especially those that migrate or fly at low altitudes, may collide with wind turbine blades, resulting in injury or death. Similarly, bats are also at risk of colliding with the rotating blades. These collisions can have significant consequences for populations of certain bird and bat species.

To mitigate this impact, wind farms are often strategically located away from known migration routes and important habitats. Moreover, Researchers and engineers are working together to develop innovative solutions such as painting the turbine blades with colors or patterns that are visible to birds to reduce the collision risk.

Disruption of Wildlife Habitat

Apart from direct collisions, wind farms can also disrupt wildlife habitat. The construction and operation of wind turbines require clearing of land or seabed, resulting in habitat fragmentation and loss. This can negatively affect wildlife populations by reducing their available habitats or fragmenting their habitats into smaller, less viable patches.

To minimize this impact, proper environmental impact assessments and habitat management plans must be implemented during the development and operation of wind energy projects. These plans should strive to preserve or restore habitats and ensure long-term ecological sustainability.

Noise Pollution and Disturbance

Another concern associated with wind turbines is noise pollution. Although modern wind turbines are designed to be relatively quiet, the rotating blades and mechanical components can still produce noise. Prolonged exposure to this noise can have an adverse effect on both wildlife and nearby communities.

For wildlife, noise pollution can disrupt their natural behavior patterns, interfere with communication, and cause stress. In some cases, animals may even avoid areas near wind farms, further impacting their already limited available habitats.

For local communities, noise pollution from wind turbines may cause annoyance, sleep disturbance, and decreased quality of life. It is important for wind farms to be sited at appropriate distances from residential areas and to comply with noise regulations to minimize these impacts.

Visual Pollution

The visual impact of wind turbines is another concern that is often raised, particularly in scenic and natural areas. Some people find the presence of wind turbines to be visually intrusive, affecting the aesthetics of the landscape and potentially impacting tourism and recreation.

To address this concern, careful consideration should be given to the siting and design of wind turbines. Strategic placement can help minimize the visual impact, blending them with the surrounding landscape or using technologies that make them less obtrusive.

However, it is worth noting that beauty is subjective, and what may be visually unappealing to one person may be seen as a symbol of progress and environmental responsibility to another.

Conclusion

While wind turbines offer significant benefits as a renewable energy source, it is important to acknowledge the challenges and environmental concerns associated with their deployment. Intermittency and energy storage, wildlife impact, disruption of habitat, noise pollution, and visual pollution are all factors that must be carefully considered and addressed.

By recognizing these challenges and working towards innovative solutions, the wind energy industry can continue to grow while minimizing its environmental footprint. It requires collaboration among policymakers, engineers, researchers, and communities to strike a balance between harnessing the power of wind and protecting the natural world we depend on.

International Offshore Wind Projects

Offshore wind energy has gained significant attention and momentum in recent years as countries around the world strive to reduce their reliance on fossil fuels and mitigate the impacts of climate change. While onshore wind farms have proven to be effective in generating clean energy, offshore wind projects offer unique advantages and opportunities. In this section, we will explore some of the notable international offshore wind projects and the challenges they face.

Europe's Pioneering Efforts

Europe has been at the forefront of offshore wind development, with several countries leading the way in harnessing the power of wind at sea. One of the most remarkable projects is the Horns Rev 3 offshore wind farm, located off the coast of Denmark. With a capacity of 407 MW, it is one of the largest offshore wind farms in the world. The project comprises 49 turbines, each standing at a towering height of 187 meters. It is estimated that Horns Rev 3 will provide clean energy to approximately 425,000 households, significantly reducing carbon emissions.

Another notable European offshore wind project is the London Array, situated in the Thames Estuary in the United Kingdom. It is currently the world's largest operational offshore wind farm, boasting a capacity of 630 MW. With 175 turbines spread across an area of 100 square kilometers, the London Array is an impressive example of effective utilization of offshore wind resources. It is expected to generate enough electricity to power nearly half a million homes each year.

These projects, along with many others in Europe, have not only demonstrated the technical feasibility and commercial viability of offshore wind energy but also highlighted the immense potential for job creation and economic growth.

The United States' Ambitious Goals

While Europe has taken the lead in offshore wind deployment, the United States is making significant strides in catching up. The country has vast untapped wind resources off its coastlines, providing a promising opportunity for clean energy generation. One of the pioneering projects in the U.S. is the Block Island Wind Farm, located off the coast of Rhode Island. It was the country's first commercial-scale offshore wind farm, with a modest capacity of 30 MW. Though relatively small compared to its European counterparts, the Block Island Wind Farm serves as a blueprint for future offshore wind developments in the country.

The United States is currently working towards ambitious goals to accelerate offshore wind deployment. The Biden administration has set a target of deploying 30 gigawatts (GW) of offshore wind capacity by 2030. This objective is expected to create thousands of jobs, revitalize coastal communities, and contribute to the country's clean energy transition. Projects like Vineyard Wind, off the coast of Massachusetts, and South Fork Wind, off the coast of New York, are crucial steps towards achieving these targets.

Challenges and Solutions

While offshore wind projects offer immense promise, they also come with unique challenges that need to be addressed. One significant obstacle is the higher upfront cost compared to onshore wind farms. Developing and implementing projects at sea involves substantial engineering efforts, including the construction of robust and resilient foundations to support towering wind turbines in harsh marine environments. However, as offshore wind technology matures and economies of scale are achieved, the costs are expected to decline, making offshore wind energy more economically competitive.

Another challenge is related to the environmental impacts of offshore wind farms. The construction and operation of these projects can potentially disturb marine habitats and affect marine wildlife, including fish, mammals, and birds. However, thorough environmental impact assessments and proper mitigation measures can help minimize these effects. For example, conducting pre-construction surveys to identify sensitive areas and implementing measures to reduce underwater noise during pile driving can safeguard marine life.

Furthermore, the intermittency of wind power poses a challenge for offshore wind projects. Utilizing energy storage systems, such as advanced batteries or pumped hydro storage, can help mitigate the intermittency issue and ensure a stable and reliable supply of electricity even during lulls in wind activity.

Conclusion

International offshore wind projects exemplify the potential of harnessing wind energy at sea to generate clean electricity and combat climate change. Despite the challenges associated with upfront costs, environmental impacts, and intermittency, offshore wind has proven to be a viable and valuable renewable energy source. As technology continues to advance, costs decrease, and governments commit to ambitious renewable energy targets, offshore wind is set to play a crucial role in the global energy transition. It is an exciting field that promises both environmental benefits and economic opportunities, making it a key sector to watch in the coming years.

Wind Energy vs Other Renewable Sources

Comparison with Solar Energy

Solar energy is one of the most well-known and widely used forms of renewable energy. It harnesses the power of the sun to generate electricity through photovoltaic (PV) systems. In this section, we will compare wind turbines to solar energy, exploring their similarities, differences, and the advantages and disadvantages of each.

Similarities between Wind Turbines and Solar Energy

Both wind turbines and solar energy are clean and renewable sources of power. They produce electricity without emitting greenhouse gases or contributing to air pollution. Additionally, both technologies have experienced significant advancements in recent years, making them more efficient and cost-effective.

Both wind and solar energy are variable, meaning they depend on external conditions for optimal power generation. While wind turbines rely on strong and consistent wind speeds, solar energy requires direct sunlight. To maximize their effectiveness, both wind turbines and solar panels need to be installed in locations that have constant access to their respective energy sources.

Differences between Wind Turbines and Solar Energy

One key difference between wind turbines and solar energy is the source of power they harness. Wind turbines harness the kinetic energy of the wind, while solar energy converts sunlight into electricity. This distinction means that wind turbines are more dependent on weather conditions, requiring sustained winds to generate power. Solar energy, on the other hand, can be harnessed even on cloudy days, as long as there is sufficient sunlight.

Another difference is the type of infrastructure required for each technology. Wind turbines need a designated space where the towers and turbines can be installed. This often requires large tracts of land or offshore locations. Solar energy, on the other hand, can be installed on rooftops, buildings, or in solar farms. It offers more flexibility in terms of space utilization.

Advantages of Solar Energy

Solar energy has several advantages over wind turbines. Firstly, solar panels have a longer lifespan compared to wind turbines. With proper maintenance, solar panels can last up to 30 years or more, whereas wind turbines typically have a lifespan of 20-25 years. This longevity can result in lower maintenance costs and a higher return on investment over time.

Another advantage of solar energy is its scalability. Solar panels can be installed on various scales, from small residential systems to large-scale solar farms. This adaptability makes solar energy accessible to a wider range of users, including individual homeowners, businesses, and communities.

Additionally, solar energy is more visually appealing compared to wind turbines. Solar panels can be seamlessly integrated into buildings or installed on rooftops, blending with the surroundings. In contrast, wind turbines are often tall and conspicuous, potentially impacting the aesthetic value of an area.

Disadvantages of Solar Energy

Despite its advantages, solar energy does have some drawbacks. One major challenge is its intermittent nature. Solar power generation is dependent on daylight hours and weather conditions, which can affect its reliability. Cloudy days or nighttime result in reduced or no power generation. This intermittency requires the use of energy storage systems, such as batteries, to ensure a constant power supply.

Another disadvantage is the need for a large surface area to install solar panels. Solar energy has a lower power density compared to wind energy, meaning it requires more space to generate the same amount of electricity. This can be a limiting factor, especially in densely populated areas where land availability is limited.

Moreover, the production of solar panels involves the use of rare earth metals and other materials that can have significant environmental impacts. The mining and manufacturing processes can contribute to pollution and the release of greenhouse gases. However, advancements in recycling and sustainable production practices are mitigating these concerns.

Example: Solar Energy Integration

To further illustrate the use of solar energy, let's consider an example of integrating solar panels into a residential setting.

Imagine a family living in a suburban area looking to reduce their carbon footprint and decrease their electricity bills. They decide to install solar panels on their rooftop. By conducting a thorough energy assessment of their home, they determine their energy consumption patterns and install a solar panel system that meets their needs. With the solar panels in place, they can generate a significant portion of their electricity from the sun.

To maximize their solar energy usage, they utilize energy-efficient appliances and practice energy conservation measures. They also install a battery storage system to store excess solar energy generated during the day for use during nighttime or on cloudy days. As a result, they reduce their dependence on the grid, lower their electricity bills, and contribute to a cleaner environment.

Resources and Further Reading

For further reading on solar energy and its comparison to other renewable energy sources, consider the following resources:

- "Solar Energy: The Physics and Engineering of Photovoltaic Conversion, Technologies and Systems" by Olindo Isabella, Klaus Jäger, and Arno Smets.

- "Renewable Energy: Power for a Sustainable Future" by Godfrey Boyle.

- "Renewable and Efficient Electric Power Systems" by Gilbert M. Masters.

- "Introduction to Renewable Energy" by Vaughn C. Nelson.

These resources provide in-depth knowledge and practical insights into the field of solar energy and its applications.

In conclusion, solar energy offers a promising alternative to wind turbines for generating clean and renewable electricity. While it has its limitations, such as intermittent power generation and the need for sufficient installation space, solar energy's advantages, including longer lifespan, scalability, and visual appeal, make it an attractive option for individuals, businesses, and communities seeking to embrace sustainable energy solutions. By harnessing the power of the sun, we can pave the way for a greener future.

Comparison with Hydroelectric Power

When it comes to renewable energy sources, hydroelectric power has long been recognized as a reliable and efficient method of generating electricity. In this section, we will compare the features and benefits of hydroelectric power with wind turbines to understand their respective advantages and limitations.

Overview of Hydroelectric Power

Hydroelectric power harnesses the energy of flowing or falling water to generate electricity. It involves the construction of dams, reservoirs, and powerhouses, which are equipped with turbines that convert the kinetic energy of the water into mechanical energy. This mechanical energy is then converted into electrical energy by generators.

Hydroelectric power has a rich historical background, dating back to ancient times when water wheels were used for grinding grain and powering various mechanical devices. Today, large-scale hydroelectric power plants contribute significantly to the global electricity supply.

Environmental Impact

One of the main advantages of hydroelectric power is its minimal environmental impact compared to other energy sources. Unlike wind turbines, which can have significant effects on wildlife and landscapes, hydroelectric power plants have fewer detrimental effects on ecosystems.

However, the construction of dams and reservoirs for hydroelectric projects can still lead to habitat destruction and the displacement of wildlife. The altered flow and temperature patterns of rivers can also impact fish populations and other aquatic organisms. Additionally, the decommissioning of old or inefficient hydroelectric plants can pose challenges in terms of managing the resulting waste and restoring river ecosystems.

Reliability and Capacity Factor

Hydroelectric power plants offer a high capacity factor, meaning they can consistently generate electricity at or near their maximum capacity. Since they rely on a continuous flow of water, hydroelectric plants are not subject to the intermittency issues that wind turbines face. This makes hydroelectric power a more reliable source of energy, especially for baseload electricity demand.

Wind turbines, on the other hand, depend on wind availability, which can vary greatly. While wind power has made significant advancements in terms of predicting wind patterns and optimizing turbine design, it still remains less predictable and reliable compared to hydroelectric power.

Costs and Energy Production

When it comes to cost considerations, hydroelectric power plants generally require higher upfront investment due to the construction of dams and infrastructure. However, they have lower operational and maintenance costs compared to wind turbines. The lifespan of hydroelectric power plants is also longer, typically lasting for several decades.

In terms of energy production, both hydroelectric power and wind turbines have the potential to generate significant amounts of electricity. However, the energy production of a hydroelectric plant is more predictable and controllable, as it depends on water flow and the capacity of the turbines. Wind turbines, on the other hand, are subject to variations in wind speed and direction, which can affect their energy output.

Geographical Constraints

Hydroelectric power plants require specific geographical conditions to be viable, mainly an abundance of water resources. This limits their potential locations to regions with suitable topography and hydrological characteristics. The construction of dams and reservoirs can also have social and environmental implications, as they often involve the displacement of local communities and the alteration of river ecosystems.

In contrast, wind turbines can be installed in a wider range of locations, making them more geographically flexible. However, wind resources need to meet certain criteria to be economically feasible, such as consistent wind speeds and minimal turbulence. Areas with low wind resources may not be suitable for wind power generation.

Integration and System Stability

Hydroelectric power can be easily integrated into existing electricity grid systems, as it provides a stable and controllable source of energy. The use of hydroelectric power plants for regulating grid frequency and addressing peak demand makes them valuable for overall system stability.

Wind turbines, on the other hand, introduce challenges to grid integration due to their intermittent nature. The variability and fluctuations in wind power production require additional measures for grid stability, such as energy storage systems and backup power sources. However, advancements in grid management and storage technologies are helping to improve the integration of wind power into existing systems.

Conclusion

In conclusion, hydroelectric power and wind turbines have their distinct advantages and limitations. Hydroelectric power offers a reliable and controllable source of energy with minimal environmental impact. It has a high capacity factor, long lifespan, and relatively low operational costs. On the other hand, wind turbines provide geographical flexibility, have a lower upfront investment, and contribute to diversifying the renewable energy portfolio.

Ultimately, the choice between hydroelectric power and wind turbines depends on various factors, including geographical conditions, environmental considerations, energy demand, and grid integration requirements. Both sources play important roles in the transition to a sustainable energy future, and their complementarity can help ensure a reliable and resilient electricity supply.

Comparison with Geothermal Energy

Geothermal energy is another renewable energy source that harnesses the heat from the Earth's core to generate power. In this section, we will explore the similarities and differences between wind turbines and geothermal energy, highlighting their respective advantages and disadvantages.

Principle and Technology

While wind turbines utilize the kinetic energy of the wind to generate electricity, geothermal energy harnesses the heat energy trapped beneath the Earth's surface. This heat can be accessed through geothermal power plants, which extract steam or hot water from underground reservoirs. The steam or hot water is then used to spin turbines and generate electricity.

One advantage of geothermal energy over wind turbines is its consistent power generation. Unlike wind energy, which relies on varying wind speeds, geothermal energy provides a steady and reliable source of power. This makes geothermal energy a more predictable option for meeting baseload electricity demand.

Environmental Impact

Both wind turbines and geothermal energy have relatively minimal environmental impacts compared to fossil fuel-based power generation. However, there are some differences to consider.

Geothermal power plants can release small amounts of greenhouse gases, primarily carbon dioxide and hydrogen sulfide, during the extraction and conversion process. However, these emissions are significantly lower compared to fossil fuel-based power plants.

In contrast, wind turbines do not produce greenhouse gas emissions during operation. However, the manufacturing and installation processes of wind turbines can have some environmental impact, such as the production of materials and transportation.

Resource Availability

One significant difference between wind turbines and geothermal energy is the availability of resources. Wind energy relies on the speed and consistency of the wind, making it location-dependent. Areas with consistent and strong winds are ideal for wind power generation.

On the other hand, geothermal energy resources are limited to specific areas with high heat flow from the Earth's core. These areas, such as geothermal hotspots or tectonic plate boundaries, have suitable conditions for geothermal power generation. While not as widely available as wind energy, geothermal resources have the advantage of being more constant and predictable in terms of power output.

Scalability and Flexibility

Wind turbines offer more scalability and flexibility compared to geothermal power plants. Wind farms can vary in size, ranging from small installations for individual homes to large-scale utility projects. This variability allows wind energy to be tailored to specific energy demands.

Geothermal power plants, on the other hand, require substantial initial investments and are typically larger in scale. The development of geothermal resources often involves drilling deep into the Earth's crust, which can be costly and technically challenging.

Cost Considerations

The cost of wind turbines and geothermal power plants varies depending on several factors such as resource availability, installation location, and the scale of the project.

Wind energy generally has a lower upfront cost compared to geothermal energy. However, the variability in wind speeds can impact the reliability and efficiency of wind turbines, potentially affecting energy production and overall costs.

Geothermal power plants tend to have higher initial costs due to the need for extensive drilling and infrastructure development. However, once operational, they benefit from low fuel costs and have a longer lifespan.

Applications and Limitations

Both wind turbines and geothermal power have their own applications and limitations.

Wind energy is commonly used for large-scale electricity generation, primarily through wind farms. It is also suitable for smaller applications such as powering individual homes or small communities. However, wind turbines require open spaces and may face local opposition due to their visual impact.

Geothermal energy is well-suited for baseload electricity generation and district heating systems. Its consistent power output makes it reliable for meeting continuous energy demand. However, the siting of geothermal power plants is limited to areas with suitable geothermal resources.

Conclusion

In summary, wind turbines and geothermal energy are both viable renewable energy options with their unique characteristics. Wind turbines are more widely available and offer flexibility, while geothermal power provides a consistent and reliable source of energy. The choice between the two depends on factors such as resource availability, location, and project scale. By exploring and utilizing a combination of renewable energy sources, we can pave the way for a sustainable and greener future.

Resources:

1. Bureau of Land Management. "Geothermal." Accessed March 2022. https://www.blm.gov/energy/green-energy/geothermal

2. U.S. Department of Energy. "Wind Energy Technologies." Accessed March 2022. https://www.energy.gov/eere/wind/wind-energy-technologies

3. International Geothermal Association. "Introduction to geothermal energy." Accessed March 2022. https://www.geothermal-energy.org/

4. American Wind Energy Association. "Wind energy basics." Accessed March 2022. https://www.awea.org/wind-101/basics-of-wind-energy

Future Trends and Alternative Solutions

Technological Advances in Wind Turbines

Larger and More Efficient Turbines

As renewable energy becomes an increasingly important part of our energy landscape, the need for larger and more efficient wind turbines becomes evident. In this section, we will explore the technological advances in wind turbines that aim to maximize power production while minimizing costs and environmental impact.

The Need for Larger Turbines

One of the main advantages of wind power is the scalability of the technology. By increasing the size of wind turbines, we can harness higher wind speeds at greater heights, resulting in higher energy outputs. Larger turbines also allow for reduced installation and maintenance costs per unit of energy produced.

At the heart of a wind turbine are the rotor blades, which convert the kinetic energy of the wind into mechanical energy that drives the generator. Increasing the length of the blades increases the swept area, thereby capturing more wind energy. However, longer blades can also lead to structural challenges and increased loads.

To overcome these challenges, engineers have developed advanced materials and manufacturing techniques. For example, carbon fiber composites are now used in the construction of rotor blades due to their high strength-to-weight ratio. Additionally, innovations in aerodynamics and blade design have led to improved efficiency and reduced noise levels.

Design Considerations and Challenges

Designing larger wind turbines involves several considerations and challenges. One of the primary concerns is ensuring the structural integrity of the turbine under varying wind conditions. Turbine manufacturers utilize advanced simulation software and real-world data to optimize the design and ensure safe operation.

Another challenge is transporting and erecting these massive structures. Longer blades and taller towers require specialized transportation and construction equipment. Additionally, the logistical constraints of shipping oversized components can significantly impact project timelines and costs.

Moreover, the increased size of wind turbines poses challenges for maintenance and repair. Components located at greater heights are more difficult to access, requiring specialized equipment and skilled technicians. To address this, some manufacturers are incorporating advanced sensor technology and condition monitoring systems to optimize maintenance schedules and minimize downtime.

Aerodynamic Advances

Improving the aerodynamic performance of wind turbines is crucial for maximizing energy production. This involves reducing drag, minimizing turbulence, and optimizing the airflow over the rotor blades.

One approach is the use of serrated trailing edges, which create small vortices that reduce drag and improve lift-to-drag ratios. This design innovation has shown promising results, increasing the overall efficiency of wind turbines.

Another aerodynamic advancement is the implementation of smart blade technology. Smart blades incorporate sensors and actuators that continuously monitor and adjust the shape and angle of the blades to optimize performance in real-time. This allows turbines to adapt to changing wind conditions, resulting in higher energy capture.

The Future of Large Turbines

The development of larger and more efficient wind turbines is an ongoing process driven by the need for increased renewable energy generation. Exciting advancements on the horizon include the exploration of even taller towers and the utilization of offshore locations with stronger and more consistent winds.

Additionally, the integration of energy storage systems with large wind farms will enable a more stable and reliable power supply. Excess energy generated during low demand periods can be stored and used during peak demand, reducing reliance on traditional fossil fuel-based power plants.

As the wind energy industry continues to grow, innovations in turbine design and technology will drive the efficiency and cost-effectiveness of wind power. With these advancements, wind turbines will not only become more environmentally friendly but also economically competitive with conventional energy sources.

Summary

In this section, we have explored the advancements in larger and more efficient wind turbines. Through the use of advanced materials, improved aerodynamic design, and technological innovations, wind turbines are becoming more capable of harnessing the power of the wind. These larger turbines provide increased energy output, reduced costs, and a more sustainable form of energy generation. The future of wind power looks promising as we continue to push the boundaries of turbine size and efficiency, bringing us one step closer to a greener and more sustainable future.

Floating Wind Turbines

Floating wind turbines are a fascinating and innovative solution to the challenges posed by traditional fixed-bottom offshore wind installations. With the ability to harness the strong and consistent winds that occur further offshore, floating wind turbines offer great potential for the future of renewable energy.

Background

Fixed-bottom offshore wind turbines are typically installed in shallow waters, where the turbine structure is attached or embedded into the seabed. However, this limits their potential locations to areas with suitable seabed conditions and restricts their deployment to water depths of up to 60 meters.

Floating wind turbines, on the other hand, are designed to be deployed in deeper waters where the seabed is unsuitable for conventional installations. By employing advanced floating platforms that are moored to the seabed with anchors or dynamic positioning systems, these turbines can operate in water depths of over 100 meters, opening up vast expanses of the ocean for wind energy generation.

Principles of Floating Wind Turbines

The principle behind floating wind turbines is similar to that of fixed-bottom offshore turbines. They convert the kinetic energy of the wind into mechanical energy through the rotation of a rotor, which in turn drives a generator to produce electricity. However, the floating platform adds an extra layer of complexity to the design and operation of these turbines.

The floating platform is designed to provide stability and support to the turbine, ensuring that it remains upright and stable even in rough sea conditions. Various types of floating platforms have been developed, including tension leg platforms, semi-submersibles, and spar buoys. These platforms are engineered to reduce motion in response to waves and currents, minimizing the impact on turbine performance and ensuring the safety of the structure.

Benefits of Floating Wind Turbines

Floating wind turbines offer several benefits over their fixed-bottom counterparts. One of the key advantages is the ability to access stronger and more consistent winds in deeper waters. The higher wind speeds and lower turbulence experienced further offshore result in higher energy yields and increased capacity factors for floating wind farms.

Additionally, floating wind farms can be located far away from the coastline, reducing visual and noise impacts on coastal communities. This opens up possibilities for offshore wind development in regions where land scarcity or aesthetic concerns impede the deployment of fixed-bottom turbines.

Moreover, floating wind turbines have the potential to leverage existing offshore oil and gas infrastructure, reducing costs and accelerating the deployment of renewable energy in offshore areas that have previously been utilized for fossil fuel extraction.

Technical Challenges

While floating wind turbines offer great promise, there are several technical challenges that need to be addressed for their widespread adoption:

Platform Stability: The stability and motion control of floating platforms are critical for turbine performance and safety. Designing platforms that can withstand extreme weather conditions, while minimizing motion and ensuring the stability of the turbine, is a significant engineering challenge.

Mooring and Anchoring Systems: The mooring and anchoring systems play a crucial role in keeping the floating turbine in position. Developing reliable and cost-effective systems that can withstand the harsh marine environment and provide secure mooring is essential for the success of floating wind farms.

Grid Connection: Floating wind farms, especially those located further offshore, face challenges in terms of electrical grid connection. Efficient and reliable transmission systems need to be developed to connect these remote wind farms to onshore grids.

Case Study: Hywind Scotland

Hywind Scotland, developed by Equinor (formerly Statoil), is one of the pioneering floating wind farms in the world. Located approximately 25 kilometers off the coast of Scotland, Hywind Scotland utilizes five floating wind turbines, each with a capacity of 6 MW.

The turbines are mounted on spar buoys, which provide stability and minimize movements in response to waves. The buoyant section of the spar is filled with ballast, keeping the turbine upright, while the mooring lines and suction anchors secure it to the seabed.

Hywind Scotland has demonstrated the feasibility of floating wind technology, with the capacity factor exceeding expectations. The project has been successful in generating clean and renewable electricity, while also gaining valuable insights into the performance and reliability of floating wind turbines.

Future Outlook

As technology continues to advance, floating wind turbines are expected to play an increasingly significant role in the global energy transition. With their ability to tap into abundant wind resources further offshore, floating wind farms have the potential to contribute significantly to the renewable energy targets of many countries.

Ongoing research and development efforts are focused on improving the reliability and cost-effectiveness of floating wind turbines. This includes advancements in platform design, mooring systems, grid integration, and turbine technologies. As these technologies mature, floating wind farms will become more economically viable and competitive with other renewable energy sources.

In conclusion, floating wind turbines offer an exciting and promising avenue for the expansion of offshore wind energy. With their ability to access deeper waters and harness stronger winds, they have the potential to deliver significant clean electricity generation while minimizing the visual and environmental impacts associated with conventional offshore wind turbines. Through continued innovation and investment, floating wind technology can drive the growth of renewable energy and contribute to a more sustainable future.

Vertical Axis Wind Turbines

Vertical axis wind turbines (VAWTs) are a type of wind turbine where the main rotor shaft is set vertically. Unlike traditional horizontal axis wind turbines (HAWTs), which have blades that rotate on a horizontal plane like a propeller, VAWTs have blades that rotate around a vertical axis resembling an eggbeater. This unique design offers several advantages and disadvantages compared to HAWTs, making them an interesting alternative solution in the field of wind energy.

Principles of VAWTs

To understand how VAWTs work, let's delve into the principles behind their operation. VAWTs rely on the aerodynamic forces produced by the wind interacting with their blades to generate rotational motion. When the wind flows over the blades, it creates a difference in air pressure between the windward and leeward sides. This pressure difference creates lift and drag forces on the blades, causing them to rotate.

One key advantage of VAWTs is their ability to capture wind from any direction. As the blades are mounted vertically, they can effectively harness wind from multiple directions without needing to be constantly repositioned. This makes VAWTs suitable for urban environments and areas with turbulent wind patterns.

Types of VAWTs

There are several different designs of VAWTs, each with its own set of characteristics and advantages. Three common types are the Savonius, Darrieus, and Giromill.

1. **Savonius VAWTs** have a simple design, consisting of curved blades that resemble half-cylinders. They are self-starting and can generate power at low wind speeds. However, they have low efficiency and generate less power compared to other VAWT designs.

2. **Darrieus VAWTs** are characterized by their vertical airfoils or blades that are arranged in an eggbeater shape. This design offers higher efficiency and can generate power at higher wind speeds. However, Darrieus VAWTs require external means to start rotating, such as motors or additional wind sources, due to their lack of self-starting capability.

3. **Giromill VAWTs**, also known as helical or H-rotor VAWTs, have straight blades that resemble a helix or the letter "S". This design offers a good compromise between the simplicity of the Savonius and the efficiency of the Darrieus. Giromill VAWTs are self-starting, have good power output, and can operate at varying wind speeds.

Advantages of VAWTs

VAWTs offer several advantages over their HAWT counterparts:

- **Ease of installation:** VAWTs can be installed on lower towers, making them easier and more cost-effective to install compared to HAWTs, which require taller towers for sufficient wind exposure.

- **Omni-directional wind capture:** As mentioned earlier, the vertical axis allows VAWTs to capture wind from any direction, eliminating the need for complex tracking mechanisms that HAWTs require.

- **Less noise and vibration:** VAWTs produce less noise and vibration compared to HAWTs, making them more suitable for urban and residential areas where noise pollution is a concern.

- **Less susceptibility to strong gusts:** VAWTs have a higher tolerance for strong gusts of wind due to their symmetric design. This makes them more reliable and less prone to damage during extreme weather events.

- **Ease of maintenance:** VAWTs have a simpler design and fewer moving parts, making them easier to maintain and repair. This reduces maintenance costs and improves overall reliability.

Disadvantages of VAWTs

Despite their advantages, VAWTs also have some drawbacks:

- **Lower efficiency:** VAWTs generally have lower efficiency compared to HAWTs. This is partly due to the inherent drag forces generated by the vertical rotor, which can decrease power output.

- **Higher start-up torque:** VAWTs require a higher start-up torque compared to HAWTs. This means that additional external sources, such as motors, may be necessary to initiate rotation in low wind conditions.

- **Tower height limitations:** VAWTs are not ideal for capturing wind at higher altitudes, as their vertical axis design requires them to be closer to the ground. This limits their exposure to high-speed winds available at elevated heights.

- **Complexity in electricity generation:** The rotational motion of VAWTs needs to be converted into usable electricity through a series of mechanical and electrical components. This complexity can lead to additional costs and maintenance requirements.

Future Trends and Alternative Solutions

The field of VAWTs is continually evolving, with ongoing research and development aimed at improving their efficiency and addressing their limitations. Some future trends and alternative solutions include:

- **Advanced blade designs:** Researchers are exploring various blade designs, materials, and configurations to enhance the aerodynamic performance of VAWTs. These advancements could significantly improve their efficiency and power output.

- **Combined VAWT-HAWT systems:** Combining the strengths of VAWTs and HAWTs into hybrid systems shows promise. By utilizing both vertical and horizontal axis turbines together, the hybrid systems can optimize energy extractio=n and adaptability to varying wind conditions.

- **Integration with energy storage systems:** Pairing VAWTs with energy storage systems, such as batteries or pumped hydro storage, can help overcome the intermittency of wind energy. This integration allows for a more reliable and consistent power supply.

- **Micro VAWTs for localized power generation:** Miniaturized VAWTs, known as micro VAWTs, are being developed for localized power generation scenarios. These smaller-scale turbines can be installed in urban areas, on rooftops, or in remote off-grid locations, providing clean energy solutions.

- **Combined renewable energy systems:** Integrating VAWTs with other renewable energy sources, such as solar photovoltaic systems or geothermal energy, can create hybrid systems that maximize energy generation from multiple sources. These integrated systems can improve overall power reliability and reduce reliance on a single energy source.

While VAWTs may not currently dominate the wind energy landscape, their unique advantages and ongoing advancements make them an exciting area of research and development. As we strive for a sustainable and renewable future, exploring alternative solutions like VAWTs can contribute to a cleaner and greener energy mix. Whether they become the "swam ass fans" or the future of wind energy, only time will tell.

Other Renewable Energy Sources

Solar Photovoltaic Systems

In this section, we will explore solar photovoltaic systems, which are a key component of renewable energy. Photovoltaic (PV) systems use solar panels to convert sunlight directly into electricity. These systems have gained significant attention in recent years due to their potential to reduce greenhouse gas emissions and provide a sustainable energy source.

Background

Solar energy is a vast and inexhaustible resource. The sun provides an abundance of energy to Earth every day, and harnessing it through solar PV systems is a promising solution to meet our energy needs. Solar panels, or PV modules, consist of semiconductor materials such as silicon, which generate an electric current when exposed to sunlight.

Principles of Solar Photovoltaic Systems

The principles behind solar photovoltaic systems are based on the photovoltaic effect. This effect occurs when photons from sunlight hit the PV module and transfer their energy to the semiconductor material. This energy causes the electrons in the material to become excited and flow, generating an electric current.

Components of a Solar Photovoltaic System

A solar photovoltaic system consists of several key components:

1. **Solar Panels:** These are the main components of the system and are made up of interconnected PV modules. They capture sunlight and convert it into electricity.

2. **Inverter:** The inverter is responsible for converting the direct current (DC) generated by the solar panels into alternating current (AC) that can be used to power electrical devices in homes, businesses, or the grid.

3. **Battery Storage (Optional):** In some solar PV systems, batteries are used to store excess electricity generated during the day for use during periods of low sunlight or high energy demand.

4. **Mounting and Tracking Systems:** Solar panels need to be mounted securely, either on rooftops or on the ground. Tracking systems can be used to optimize the angle and orientation of the panels to maximize sunlight absorption.

5. **Balance of System (BOS) Components:** BOS components include electrical wiring, switches, and protective devices necessary to connect the solar panels to the electrical grid or the load.

Efficiency and Performance

The efficiency of a solar photovoltaic system refers to the percentage of sunlight that is converted into usable electricity. Several factors affect the efficiency of solar panels, including the quality of the semiconductor materials, management of sunlight absorption, and the operating temperature of the panels.

Advancements in technology have led to increased efficiency of solar panels. While conventional silicon-based solar panels have an average efficiency of 15-20

Integration and Grid Connection

Solar PV systems can be installed in a variety of settings, ranging from residential rooftops to large-scale solar farms. Integrating these systems with the electrical grid allows the electricity generated to be fed into the existing infrastructure, supplying power to homes, businesses, and other consumers.

To ensure seamless integration, grid-interactive inverters are used. These inverters synchronize the generated electricity with the voltage and frequency of the grid, allowing excess power to be exported and drawing power from the grid when needed. This integration facilitates the widespread adoption of solar PV systems and promotes a more sustainable energy mix.

Advantages and Benefits

Solar photovoltaic systems offer numerous advantages and benefits:

- **Clean and Renewable:** Solar energy is clean and does not produce harmful emissions or contribute to climate change. It is a sustainable energy source that reduces our reliance on fossil fuels.

- **Energy Cost Savings:** Installing solar panels can lead to significant cost savings on electricity bills over the system's lifetime. As solar resources are free, the main costs are associated with the initial installation and maintenance.

- **Energy Independence:** Solar PV systems provide energy independence by generating electricity on-site. This reduces dependence on external energy sources and mitigates the impact of power outages.

- **Job Creation:** The growing solar industry creates employment opportunities in manufacturing, installation, and maintenance of solar PV systems, contributing to economic growth.

- **Scalability:** Solar PV systems can be installed at various scales, from small residential systems to large utility-scale projects, making them customizable to meet diverse energy needs.

Challenges and Considerations

While solar photovoltaic systems offer numerous benefits, certain challenges and considerations need to be taken into account:

- **Intermittency:** Solar energy production is dependent on sunlight, meaning it is intermittent and not available at night or during periods of low sunlight. Battery storage and grid integration help overcome this challenge.

- **Space Requirements:** PV systems require significant space, especially for large-scale installations. Balancing the land requirements with environmental considerations and competing land uses can be a challenge.

- **Upfront Costs:** The initial installation costs of solar PV systems can be a barrier to adoption for some individuals or businesses. However, the falling costs of solar panels and available incentives, such as tax credits, can help offset these costs.

- **Maintenance and Durability:** Regular maintenance is required to ensure optimal performance of solar panels. Additionally, the lifespan of solar panels can vary, typically ranging from 25 to 30 years, depending on the quality and technology used.

Conclusion

Solar photovoltaic systems hold tremendous potential as a clean and sustainable energy solution. The ever-improving efficiency and decreasing costs of solar panels, along with advancements in storage and grid integration technologies, are making solar PV systems increasingly viable.

While challenges remain, the benefits of solar photovoltaic systems outweigh the drawbacks. As we strive to transition to a more sustainable energy future, embracing solar PV technology can contribute significantly to reducing greenhouse gas emissions, enhancing energy security, and promoting economic growth.

Remember, the sun is not just a celestial ball of gas; it has become a mighty ally in our fight against the eleventh plague of our times—the climate crisis. So, let us harness the power of the sun, one photon at a time!

Biomass Energy

Biomass energy is a form of renewable energy derived from organic matter such as plants, agricultural residues, and animal waste. In this section, we will explore the principles, benefits, challenges, and future prospects of biomass energy.

Principles of Biomass Energy

The principle behind biomass energy lies in harnessing the chemical energy stored in organic materials through the process of combustion or decomposition. When biomass is burned, heat is released, which can be used to generate steam and drive a turbine to produce electricity. Alternatively, biomass can be converted into biogas through anaerobic digestion or fermented into bioethanol and biodiesel.

The energy content of biomass is derived from photosynthesis, the natural process by which plants convert sunlight, carbon dioxide, and water into organic matter. This process captures carbon dioxide from the atmosphere, making biomass a carbon-neutral energy source since the carbon released during combustion is equal to the carbon absorbed during plant growth.

Benefits of Biomass Energy

1. Renewable and Sustainable: Biomass energy relies on organic materials that can be continuously replenished through agricultural and forestry practices. Unlike fossil fuels, biomass is a sustainable and readily available source of energy.

2. Reduction of Greenhouse Gas Emissions: Biomass has the potential to reduce greenhouse gas emissions compared to fossil fuels. While the combustion of biomass releases carbon dioxide, the growth of biomass plants can absorb an equivalent amount of carbon dioxide, resulting in a net-zero carbon footprint.

3. Waste Management Solution: Biomass energy can utilize organic waste materials that would otherwise contribute to pollution, such as agricultural residues, food scraps, and animal manure. By converting these waste materials into energy, biomass helps reduce waste and landfill emissions.

4. Versatility: Biomass energy can be used for various applications, including electricity generation, heating, and transportation. It can be co-fired with fossil fuels in existing power plants, providing a cost-effective and efficient transition to renewable energy.

Challenges and Considerations

1. Feedstock Availability and Logistics: The availability and consistent supply of biomass feedstock can be a challenge. It requires careful planning and management of feedstock sources, including the establishment of sustainable biomass plantations and efficient transportation systems.

2. Environmental Impacts: The cultivation and harvesting of biomass feedstock can have environmental consequences, such as deforestation, soil degradation, and water pollution. Sustainable practices, including responsible land management and cultivation techniques, need to be implemented to mitigate these impacts.

3. Air Pollution: While biomass energy is considered carbon-neutral, the combustion process can release air pollutants such as particulate matter, nitrogen oxides, and volatile organic compounds. Proper emission control technologies and monitoring systems are necessary to minimize air pollution.

4. Technological Challenges: Biomass energy technologies require continuous research and development to improve efficiency and address technical challenges, such as the high moisture content of some biomass feedstock, variability in feedstock composition, and the integration of biomass energy into existing power infrastructure.

Examples and Case Studies

1. Biogas from Agricultural Waste: In rural areas, agricultural waste such as crop residues and animal manure can be anaerobically digested to produce biogas, which can be used for cooking, lighting, and electricity generation. This approach not only offers energy access but also provides an environmentally friendly solution for waste management.

2. Biomass Power Plants: In countries like Sweden and Finland, biomass power plants have been successfully integrated into the energy mix. These power plants utilize locally sourced biomass feedstock, including forestry residues and wood waste, to generate electricity and district heating, reducing reliance on fossil fuels and supporting rural economies.

Future Prospects and Alternative Solutions

1. Advanced Conversion Technologies: Research is underway to develop advanced biomass conversion technologies, such as gasification and pyrolysis, that can efficiently convert biomass into a range of useful products, including biofuels, biochemicals, and bio-based materials.

2. **Algae-based Biofuels:** Algae hold promise as a sustainable feedstock for biofuel production. Algae can be grown in wastewater or in cultivation systems that do not compete with agricultural land. Furthermore, algae have a high oil content, making them a potential source for biodiesel production.

3. **Integration with Carbon Capture and Storage:** Biomass energy can be coupled with carbon capture and storage technologies to achieve negative carbon emissions. By capturing and storing the carbon dioxide emitted during biomass combustion, the overall carbon footprint can be further reduced.

Conclusion

Biomass energy offers a renewable and sustainable solution to our energy needs, utilizing organic waste materials and plant-based resources to generate electricity, heat, and transportation fuels. Despite its challenges, biomass energy can play a significant role in reducing greenhouse gas emissions and diversifying our energy mix. Continued research and development in biomass conversion technologies, sustainable feedstock management, and environmental considerations are vital to unlocking the full potential of biomass energy. By embracing biomass energy as part of a comprehensive renewable energy strategy, we can move closer to a cleaner and more sustainable future.

Tidal and Wave Energy

Tidal and wave energy are forms of renewable energy that harness the power of the ocean to generate electricity. Both these sources have immense potential to contribute to our energy needs while reducing our reliance on fossil fuels. In this section, we will explore the principles and technologies behind tidal and wave energy, their advantages and challenges, and their role in the future of renewable energy.

Principles of Tidal Energy

Tidal energy is derived from the gravitational pull of the moon and the sun on the Earth's oceans. As the tides rise and fall, the flow of water can be captured and converted into electricity. Tidal power plants typically utilize two main methods for electricity generation:

1. **Tidal Barrages:** A tidal barrage is a dam-like structure built across a bay or estuary. As the tide rises, water flows through turbines in the structure, driving generators to produce electricity. During low tide, the gates are

closed, and the water stored behind the barrage is released when the tide turns. Tidal barrages are effective in harnessing large amounts of energy but can have significant environmental impacts, such as habitat disruption and altered sediment patterns.

2. **Tidal Turbines:** Tidal turbines are similar to wind turbines but are designed to operate underwater. These turbines are placed on the seabed or installed on floating platforms and use the kinetic energy of tidal currents to generate electricity. Tidal turbines are more environmentally friendly than barrages and can be positioned in deeper waters, minimizing their impact on marine life and allowing for greater flexibility in deployment.

Principles of Wave Energy

Wave energy, on the other hand, is derived from the motion of ocean waves. Waves are caused by wind blowing over the surface of the ocean, and their constant motion can be converted into electrical energy. Wave energy technologies typically fall into two main categories:

1. **Point Absorbers:** Point absorbers are buoy-like devices that float on the surface of the water. They are designed to move with the waves, capturing the vertical motion and converting it into electricity using mechanical systems, such as pistons or oscillating water columns. Point absorbers are highly scalable and can be deployed individually or in arrays, making them suitable for various locations and power requirements.

2. **Overtopping Devices:** Overtopping devices are typically shore-based structures designed to capture the energy of waves that crash onto a reservoir. As waves enter the reservoir, the water level rises, and the potential energy is then used to drive turbines, generating electricity. This method is efficient in areas with high wave energy and large tidal ranges but requires specific coastal geography for optimal performance.

Advantages and Challenges

Tidal and wave energy offer several advantages over other forms of renewable energy:

- **Predictability:** Both tidal and wave energy are highly predictable, as their sources (tides and waves) can be accurately forecasted. This predictability makes it easier to integrate their power generation into the electricity grid and plan for future energy demands.

- **Consistency:** Tidal and wave energy sources are more consistent than wind or solar power, which are influenced by weather patterns. This consistency ensures a more reliable supply of electricity, reducing the need for energy storage technologies.

- **High Energy Density:** The energy present in tidal and wave environments is significantly higher compared to wind or solar resources. This high energy density allows for the extraction of substantial amounts of power from relatively small areas, making tidal and wave energy efficient in terms of land and space utilization.

However, there are also challenges associated with tidal and wave energy:

- **High Initial Costs:** The construction of tidal barrages or deployment of wave energy devices can be expensive, requiring substantial upfront investment. These high initial costs can be a barrier to the widespread adoption of tidal and wave energy technologies.

- **Environmental Impacts:** Tidal barrages, in particular, can have significant environmental impacts due to their alteration of estuarine ecosystems and disruption of natural sedimentation patterns. The impact of tidal turbines and wave energy devices on marine life is still being studied, although they are generally considered less harmful than barrages.

Current Projects and Future Outlook

Tidal and wave energy technologies are still in the early stages of development and deployment. However, several notable projects around the world show promise for the future:

- **MeyGen Tidal Array, Scotland:** The MeyGen project, located in the Pentland Firth, Scotland, is one of the largest tidal energy schemes in the world. It features an array of tidal turbines that generate clean electricity from tidal currents. The project has the potential to power thousands of homes and pave the way for further tidal energy developments.

- **Wave Hub, Cornwall, UK:** The Wave Hub is a testing facility located off the coast of Cornwall, UK, that provides a grid-connected infrastructure for testing and demonstrating wave energy devices. It supports the deployment and testing of various wave energy technologies, helping to advance the sector as a whole.

The future outlook for tidal and wave energy is promising, with ongoing research and technological advancements driving the development of more efficient and cost-effective systems. As these technologies mature and overcome their current challenges, they have the potential to play a significant role in our transition to a cleaner and more sustainable energy future.

Energy Efficiency and Conservation

Importance of Reducing Energy Consumption

In today's modern world, energy consumption has reached unprecedented levels. Our incessant need for energy to power our homes, businesses, and industries has put a tremendous strain on the planet. This has led to a myriad of environmental issues, including climate change, air pollution, and the depletion of natural resources. It is high time we recognize the importance of reducing our energy consumption and take steps to mitigate the damaging impact of our energy-intensive lifestyles.

The Environmental Consequences

The excessive use of energy contributes significantly to greenhouse gas emissions, which are the primary cause of global warming and climate change. The burning of fossil fuels for energy generation releases vast amounts of carbon dioxide into the atmosphere, trapping heat and causing the Earth's temperature to rise. This increase in temperature leads to adverse effects such as melting ice caps, rising sea levels, extreme weather events, and disruptions in ecosystems.

Furthermore, the reliance on conventional energy sources, such as coal and natural gas, for electricity generation also leads to air pollution. Harmful pollutants like sulfur dioxide, nitrogen oxides, and particulate matter are released into the air, resulting in respiratory problems, cardiovascular diseases, and even premature deaths. By reducing energy consumption, we can reduce the need for these polluting energy sources and create a cleaner and healthier environment.

Energy Efficiency as a Solution

One of the most effective ways to reduce energy consumption is through energy efficiency. Energy efficiency refers to the use of technology and practices that require less energy to perform the same tasks. It aims to maximize output while minimizing energy input, thereby reducing waste and inefficiency. Implementing energy-efficient measures in our homes, workplaces, and industries can have significant benefits for both the environment and our wallets.

There are several ways to improve energy efficiency. First and foremost, it is essential to invest in energy-efficient appliances and equipment. Look for products with the ENERGY STAR label, which indicates that they meet strict energy efficiency standards. These appliances are designed to consume less energy without compromising on performance.

In addition to using energy-efficient products, adopting energy-saving habits can also make a difference. Simple actions such as turning off lights when not in use, using natural light during the day, unplugging electronic devices when they are not being used, and properly insulating buildings can lead to substantial energy savings. These small steps, when practiced collectively, can add up to significant reductions in energy consumption.

The Benefits of Energy Reduction

Reducing energy consumption brings numerous advantages that extend beyond environmental protection. Firstly, it saves money. By being mindful of our energy usage and implementing energy-saving measures, we can significantly lower our energy bills. This allows us to allocate our financial resources to other important aspects of our lives.

Furthermore, energy reduction promotes energy independence. By relying less on non-renewable energy sources, we can decrease our dependence on foreign oil and gas. This strengthens our national security and allows us to take control of our energy future. Investing in renewable energy technologies, such as solar panels and wind turbines, can further enhance our energy independence and resilience.

Finally, reducing energy consumption fosters innovation and job creation. The transition to a low-carbon economy requires expertise in renewable energy sources, energy-efficient technologies, and sustainable practices. This shift in focus can lead to the development of new industries, the growth of green jobs, and the stimulation of the economy.

Challenges and Solutions

While reducing energy consumption is crucial, it is not without its challenges. One of the primary obstacles is the lack of awareness and knowledge among the general public. Many people are unaware of the environmental consequences of their energy use or the potential benefits of energy reduction. Education and outreach programs are essential in raising awareness and encouraging individuals to take action.

Another challenge is the initial cost of implementing energy-efficient measures. Energy-efficient appliances and equipment, although cost-effective in the long run, may have a higher upfront price tag. However, various financial incentives, such as tax credits, rebates, and low-interest loans, are available to offset these costs. Governments at all levels should continue to support these programs and make them accessible to a broader audience.

In conclusion, reducing energy consumption is of paramount importance in addressing the environmental, social, and economic challenges we face today. By embracing energy efficiency, we can mitigate climate change, improve air quality, and conserve valuable resources. It is time for individuals, businesses, and governments to come together and make a concerted effort to reduce energy consumption. Through collective action, we can create a sustainable future for generations to come.

Energy-efficient Buildings and Appliances

In the quest for a sustainable future, energy-efficient buildings and appliances play a crucial role. They contribute to reducing energy consumption, minimizing carbon emissions, and saving money on utility bills. In this section, we will explore the principles of energy efficiency, techniques for designing energy-efficient buildings, and the latest advancements in energy-efficient appliances.

Principles of Energy Efficiency

Energy efficiency is all about optimizing energy use and reducing waste. It involves implementing strategies that enable buildings and appliances to perform their required functions while consuming the least amount of energy. The following principles are fundamental to achieving energy efficiency:

1. **Insulation:** Well-insulated buildings retain heat during cold weather and keep the interior cool during hot weather. Insulating materials such as fiberglass, cellulose, and foam help regulate temperature, reducing the need for excessive heating or cooling.

2. **Smart Design:** Architectural choices that maximize natural lighting and ventilation can minimize the reliance on artificial lighting and air conditioning. Incorporating elements like skylights, large windows, and airflow design can significantly reduce energy consumption.

3. **Efficient Appliances:** Energy-efficient appliances, including refrigerators, air conditioners, and washing machines, rely on advanced technologies to consume less electricity without compromising performance. Look for appliances with Energy Star labels to ensure their energy-saving capabilities.

4. **Renewable Energy Integration:** Integrating renewable energy sources, such as solar panels or wind turbines, into buildings can significantly reduce reliance on the grid and decrease energy consumption.

Designing Energy-efficient Buildings

Designing energy-efficient buildings requires careful consideration of multiple factors. From the layout to the choice of materials, every aspect plays a role in the overall energy performance. Here are some key considerations when designing energy-efficient buildings:

1. **Orientation and Shading:** Optimizing a building's orientation to maximize exposure to sunlight can reduce the need for artificial lighting and heating. Incorporating shading devices like awnings or overhangs can mitigate excessive heat gain during the summer months.

2. **Windows and Glazing:** Using energy-efficient windows with low-emissivity (low-e) coatings and double or triple glazing can reduce heat loss or gain. Proper window placement and insulation around frames are also critical to minimize energy wastage.

3. **Efficient HVAC Systems:** Heating, ventilation, and air conditioning (HVAC) systems account for a significant portion of a building's energy consumption. Employing high-efficiency HVAC systems with programmable thermostats, variable-speed drives, and zone controls can optimize energy usage.

4. **Lighting:** Choosing energy-efficient lighting solutions, such as LED bulbs or compact fluorescent lamps (CFLs), can drastically reduce electricity consumption for lighting purposes. Incorporating natural lighting through skylights or large windows can further enhance energy savings.

5. **Insulation and Air Sealing:** Proper insulation of walls, roofs, and floors coupled with effective air sealing techniques eliminates air leaks, prevents drafts, and ensures stable indoor temperatures, thereby reducing the need for excessive heating or cooling.

It is essential to note that obtaining energy-efficient buildings requires a holistic approach. Collaboration between architects, engineers, builders, and occupants is crucial to achieve optimal energy performance.

Advancements in Energy-efficient Appliances

Advancements in technology have led to the development of energy-efficient appliances that offer substantial energy savings without sacrificing functionality. Here are some of the latest innovations in energy-efficient appliances:

1. **Smart Thermostats:** Smart thermostats enable homeowners to control heating and cooling settings remotely. They learn user preferences and adjust temperatures accordingly, leading to optimized energy usage and cost savings.

2. **Energy-efficient Refrigerators:** Modern refrigerators are equipped with advanced features such as improved insulation, smart defrosting systems, and variable-speed compressors that reduce energy consumption while keeping food fresh and safe.

3. **High-efficiency Washing Machines:** Energy-efficient washing machines use advanced water-recirculation systems, low-water-consumption designs, and efficient spin cycles to minimize water and energy use, ultimately reducing both utility costs and environmental impact.

4. **LED Lighting:** Light-emitting diode (LED) lighting has revolutionized the lighting industry. LED bulbs consume significantly less energy and have a longer lifespan compared to conventional incandescent or fluorescent bulbs.

5. **Smart Power Strips:** Smart power strips detect power usage patterns and automatically cut off power supply to standby devices when they are not in use. This eliminates energy wastage from devices in standby mode.

While energy-efficient appliances may initially require a higher upfront investment, the long-term energy savings and environmental benefits make them worthwhile.

Resource Management and Consumer Behavior

In addition to sustainable design and energy-efficient appliances, resource management and consumer behavior have a significant impact on overall energy sustainability. Here are some tips to effectively manage resources and adopt energy-saving practices:

1. **Energy Audits:** Conduct regular energy audits to identify areas of improvement and energy-saving opportunities within buildings. These audits provide valuable insights into energy usage patterns and help prioritize efficiency upgrades.

2. **Smart Metering:** Smart metering systems provide real-time information on energy consumption, enabling users to monitor and adjust their usage

accordingly. This data empowers consumers to make informed choices regarding energy consumption.

3. **Behavioral Changes:** Encouraging energy-saving habits like turning off lights when not in use, unplugging electronics, and utilizing natural lighting can significantly reduce energy consumption. Education and awareness campaigns play a crucial role in driving positive behavioral changes.

4. **Water Conservation:** Conserving water indirectly contributes to energy efficiency as energy is required for water treatment, heating, and distribution. Implementing efficient water fixtures, rainwater harvesting systems, and water recycling methods can help conserve this valuable resource.

5. **Education and Training:** Educating building occupants about the importance of energy efficiency, providing training on energy-saving practices, and creating awareness about the environmental impact of energy consumption fosters an energy-conscious culture.

By adopting these strategies and embracing energy-efficient lifestyles, individuals and communities can significantly contribute to a sustainable future.

Unconventional Tip: Energy-Efficient Landscaping

While we primarily focus on building design and appliances, the surrounding landscape also plays a role in energy efficiency. Here's an unconventional yet relevant tip: energy-efficient landscaping.

Strategically planting trees and shrubs around buildings can provide shade during hot summer months, reducing the need for excessive air conditioning. Additionally, positioning trees to block cold winter winds can act as natural insulation, minimizing heat loss.

Moreover, implementing xeriscaping techniques, which involve using native plants and minimizing water usage for landscaping, can conserve both water and energy. Xeriscaping eliminates the need for excessive irrigation, reducing the energy required for pumping and distributing water.

By incorporating energy-efficient landscaping practices, we can enhance the overall energy performance of buildings and further contribute to a sustainable future.

Summary

In this section, we explored the principles of energy efficiency and discussed techniques for designing energy-efficient buildings. We also examined the latest advancements in energy-efficient appliances, emphasizing the importance of resource management and consumer behavior. By implementing energy-efficient practices and embracing sustainable design principles, we can mitigate energy waste, reduce carbon emissions, and create a more sustainable future for generations to come. Remember, every small step counts towards making a significant impact. So let's embrace energy efficiency and build a greener tomorrow.

Sustainable Transportation Solutions

Transportation plays a vital role in our daily lives, enabling us to commute to work, travel, and transport goods. However, traditional transportation methods heavily rely on fossil fuels, which contribute to environmental pollution and climate change. In this section, we will explore sustainable transportation solutions that aim to reduce energy consumption, minimize environmental impact, and promote a more sustainable future.

Importance of Sustainable Transportation

Transportation is a significant contributor to greenhouse gas emissions, mainly due to the combustion of fossil fuels in vehicles. These emissions contribute to air pollution, climate change, and negative health effects. Therefore, transitioning to sustainable transportation is crucial in mitigating these issues and achieving a more sustainable society. Sustainable transportation aims to reduce reliance on fossil fuels, improve energy efficiency, and minimize environmental impact through innovative technologies and practices.

Advancements in Electric Vehicles

Electric vehicles (EVs) are a promising solution to reduce greenhouse gas emissions in the transportation sector. EVs utilize electric motors and batteries, eliminating the need for traditional internal combustion engines. They produce zero tailpipe emissions, significantly reducing air pollution. Furthermore, advancements in battery technology have improved the range and charging capabilities of EVs, making them a viable option for everyday use.

One challenge associated with EVs is the limited availability of charging infrastructure. To address this issue, governments and private organizations need to invest in the development of a comprehensive charging network. Additionally, research should focus on the recycling and disposal of EV batteries to minimize environmental impact.

Promoting Public Transportation

Public transportation systems, such as buses, trams, and trains, offer an efficient and sustainable alternative to individual car ownership. By promoting the use of public transportation, we can reduce traffic congestion, decrease energy consumption, and lower greenhouse gas emissions. Investing in the expansion and improvement of public transportation networks is crucial to encourage more people to adopt this sustainable option.

Furthermore, integrating smart technology into public transportation systems can enhance efficiency and convenience. Real-time tracking systems, mobile ticketing, and automated fare collection systems streamline operations and improve the overall passenger experience.

Active Transportation: Walking and Cycling

Active transportation, including walking and cycling, not only promotes physical health but also contributes to sustainable transportation. Encouraging people to adopt walking or cycling for short-distance trips reduces the number of vehicles on the road, leading to decreased energy consumption and emissions.

To promote active transportation, cities can invest in the development of pedestrian-friendly infrastructure, such as sidewalks, crosswalks, and bike lanes. Education campaigns and incentives can also encourage individuals to choose walking or cycling as their preferred mode of transportation for short journeys.

Integrating Renewable Energy

In addition to transitioning to electric vehicles, integrating renewable energy sources into transportation systems is essential to achieve a sustainable future. Renewable energy, such as solar and wind power, can power electric charging stations and other transportation infrastructure.

To maximize the utilization of renewable energy, governments and organizations should invest in the development of smart grid technologies. These technologies enable efficient integration of renewable energy sources into transportation systems, ensuring optimized energy production and distribution.

Incentives and Policy Support

The transition to sustainable transportation requires a collaborative effort between governments, businesses, and individuals. Governments can play a crucial role in driving this change by implementing supportive policies and offering incentives.

Examples of incentives include tax credits for purchasing electric vehicles, subsidies for public transportation fares, and grants for the development of sustainable transportation infrastructure. Additionally, policies that promote mixed-use development, walkability, and cycling infrastructure can encourage individuals to adopt sustainable transportation options.

Challenges and Future Outlook

While sustainable transportation solutions offer significant potential, they also face challenges that need to be addressed. These challenges include the high upfront cost of adopting sustainable technologies, limited charging infrastructure for electric vehicles, and resistance to change from individuals and industries.

However, advancements in technology, coupled with supportive policies and increased public awareness, are driving the shift towards sustainable transportation. The future of transportation lies in the integration of electric vehicles, efficient public transportation systems, active transportation options, renewable energy, and innovative policies.

By promoting sustainable transportation, we can create healthier and cleaner communities, reduce carbon emissions, and work towards a more sustainable and resilient future for generations to come.

Conclusion

Sustainable transportation solutions present an opportunity to address the environmental challenges associated with traditional transportation methods. By embracing electric vehicles, promoting public transportation, encouraging active transportation, integrating renewable energy, and implementing supportive policies, we can reduce energy consumption, minimize emissions, and build a more sustainable society.

While challenges exist, ongoing advancements in technology and increased awareness about the importance of sustainable transportation provide hope for a greener future. It is essential for governments, businesses, and individuals to collaborate and invest in the development and adoption of sustainable transportation solutions. Together, we can create a more sustainable and resilient world for future generations.

Index

-effectiveness, 8, 113, 115

ability, 15, 43, 45, 97, 113–116
absorption, 120
abundance, 106, 119
acceptance, 13, 14, 66, 67, 73–76, 88, 93
access, 19, 25, 27, 37–39, 50, 81, 82, 96, 102, 112, 114, 115, 124
account, 20, 34, 52, 63, 65, 69, 121
action, 130, 131
activity, 9, 27, 29, 44, 97, 100
adaptability, 103
addition, 17, 24, 27, 38, 45, 48, 71, 130, 133, 136
address, 6, 10, 11, 14, 17, 19, 22, 23, 26, 34, 39, 42, 46, 55, 56, 59, 62, 63, 67, 69–71, 73–75, 81, 84, 86, 88, 90, 91, 93, 98, 112, 124, 135, 137
adherence, 41, 94
adoption, 6, 13, 114, 120, 137
advance, 45, 81, 101, 115
advancement, 16, 47, 112
advantage, 47, 103, 107, 108, 116
aerodynamic, 1, 21, 80, 112, 113, 116

aesthetic, 10, 29–31, 33, 93, 103, 114
age, 39
air, 9, 17, 19, 68, 80, 88, 102, 116, 124, 129, 131, 134, 135
airflow, 112
algae, 125
allocation, 48
alteration, 106
alternative, 19, 20, 78, 81, 84, 88, 93, 104, 118, 119, 136
aluminum, 84
amount, 59, 97, 103, 123, 131
analysis, 65, 89
angle, 112
animal, 123, 124
annoyance, 23, 68, 70, 71, 73, 88, 98
anxiety, 71
appeal, 29–31, 104
appearance, 29, 30
application, 43
approach, 14, 16, 23, 30, 32, 50, 51, 57, 69, 70, 72, 73, 80–82, 84, 112, 124, 132
approval, 66
area, 29, 31, 71, 81, 99, 103, 111, 119
argument, 30

139

arrangement, 1, 72
array, 31
aspect, 39, 67, 71, 83, 89, 91, 132
ass, 23, 119
assessment, 71, 73, 103
atmosphere, 8, 123, 129
attention, 27, 80, 94, 99
attractiveness, 32
audience, 130
authenticity, 33
availability, 17, 59, 66, 103, 105, 108, 109, 124, 135
avenue, 115
aviation, 66, 67
awareness, 130, 137
axis, 93

background, 71, 87, 105
balance, 13, 18–20, 22–24, 26, 28, 30–33, 35, 53–55, 57, 59, 64, 67, 70, 73, 89, 99
balancing, 12, 34, 59, 60, 62
ballast, 115
barotrauma, 17, 19, 27
barrier, 23
base, 96
baseload, 105, 107, 109
bat, 15, 17, 20, 22, 27, 88, 97
battery, 42–45, 63, 64, 104
battle, 26
beauty, 28–31, 88, 98
behavior, 12, 20, 60, 81, 98, 133, 135
being, 21–23, 41, 43, 57, 68–71, 73, 85, 86, 108, 130
benefit, 75, 109
bike, 136
bio, 124
biodiesel, 123, 125
biodiversity, 16, 25, 28

bioethanol, 123
biofuel, 125
biogas, 123, 124
biomass, 123–125
bird, 16, 19, 20, 22, 26, 32, 88, 97
birdsong, 72
birdwatching, 29, 32
blackout, 61
blade, 27, 32, 82, 111, 112
bleeding, 19
blood, 71
blueprint, 100
bottom, 113, 114
breeding, 15, 16, 19, 25, 27
brink, 48
bubble, 27
budget, 40
buffer, 43
building, 13, 37, 38, 66, 94, 134
buildup, 80
burden, 47, 48
bureaucracy, 66
burning, 8, 129

cable, 24
camping, 29, 32
capacity, 1, 43, 48, 60, 61, 66, 97, 99, 100, 105–107, 114, 115
Cape Cod, 31, 92
Cape Wind, 93
capital, 43, 48
capture, 1, 9, 77, 79, 112, 116, 125
car, 136
carbon, 8, 9, 75, 87, 88, 99, 103, 107, 111, 123–125, 129–131, 135, 137
case, 34, 52, 69, 75, 84, 89, 93, 94
cast, 38
catalyst, 47

Index

causal, 13
cause, 12, 16, 17, 19, 20, 22, 24, 27, 60, 71, 98, 129
cement, 84
century, 87
chain, 45
challenge, 12, 13, 20, 48, 59, 60, 64, 81, 90, 97, 100, 103, 112, 124, 130, 135
chance, 15
change, 1, 6, 8–10, 26, 44, 49, 86–88, 99, 101, 129, 131, 135, 137
character, 29, 31
charging, 135–137
Charles F. Brush, 87
charm, 29
choice, 38, 107, 109, 132
claim, 88
cleaning, 39, 80, 81
clearing, 17, 19, 27, 37, 98
climate, 1, 6, 8–10, 26, 44, 49, 86–88, 99, 101, 129, 131, 135
clothing, 80
coal, 8, 46, 129
coast, 31, 92, 99, 100
coastline, 114
coexistence, 26
collaboration, 17, 28, 63, 84–86, 91, 99
collection, 65, 136
collision, 15, 18, 22, 24, 25, 32, 97
combat, 26, 44, 88, 101
combination, 18, 61, 109
combustion, 123–125, 135
commitment, 76, 93
communication, 12, 19, 20, 23, 24, 27, 55, 66, 67, 90–92, 98

community, 2, 23, 25, 33, 46, 52, 55, 57, 67, 69, 70, 75, 81, 88, 90, 92, 94
comparison, 104
compensation, 53, 55–57, 88, 90, 92
competition, 50
complementarity, 107
complexity, 20, 81, 91, 114
compliance, 23, 38, 69, 92
component, 9, 10, 13, 40, 76–79, 81, 94
composite, 84
composition, 25, 124
compromise, 31, 34
concept, 72
concern, 13, 15, 20, 50, 70, 71, 88, 98
conclusion, 49, 57, 82, 96, 104, 107, 115, 131
condition, 19, 82, 112
conditioning, 134
conduct, 39, 45, 75
congestion, 136
connection, 39, 66, 91, 96
connectivity, 67
consensus, 8, 93
conservation, 19, 20, 25, 28, 104
consideration, 33, 34, 43, 53, 70, 72, 75, 98, 132
consistency, 59, 108
constraint, 65, 66
construction, 9, 12, 13, 17, 19, 20, 24, 27, 43, 45, 46, 60, 67, 84, 98, 100, 105, 106, 111, 112
consultation, 10
consulting, 46
consumer, 133, 135

consumption, 103, 129–131, 135–137
contact, 16
content, 123–125
contention, 88
context, 8, 33, 88
contract, 91
contrast, 29, 103, 106, 108
contribution, 75
contributor, 135
control, 45, 60, 61, 69, 81, 124, 130
convenience, 136
conversion, 108, 124, 125
cooking, 124
cooling, 79
coordination, 66, 80
copper, 84
core, 107, 108
corrosion, 78, 80
cost, 8, 37–40, 42–44, 50, 82, 91, 93, 97, 100, 102, 106, 108, 113, 115, 123, 128, 130, 137
counsel, 55, 91
counseling, 69
country, 48, 100
creation, 45–47, 49, 75, 99, 130
crisis, 87
criticality, 81
criticism, 47, 48, 93
crop, 124
crust, 108
cultivation, 124, 125
current, 119, 128
curtailment, 27
cycle, 43
cycling, 136, 137

damage, 19, 40, 42

Darrieus, 116
date, 90
day, 42, 104, 119, 130
death, 15, 19, 24, 27, 97
debate, 13, 47, 48, 88
decision, 30, 55, 75, 90
decline, 29, 32, 69, 100
decommissioning, 13, 82–86, 105
decomposition, 123
decrease, 39, 45, 53, 61, 67, 68, 88, 101, 103, 130, 136
defense, 66
deforestation, 124
degradation, 8, 77, 124
demand, 9, 12, 42–44, 46, 50, 53, 59–62, 64, 87, 97, 105–107, 109, 112
Denmark, 84, 99
density, 43, 103
dependence, 1, 8, 88, 104, 130
dependency, 46
depletion, 129
deployment, 10, 13, 25, 42, 44, 45, 48, 59, 61, 88, 94, 99, 100, 113, 114, 127
depreciation, 48
depth, 104
design, 4, 23, 27, 30–32, 45, 66, 68, 72, 73, 78, 87, 93, 98, 105, 111–115, 133–135
designing, 131, 132, 135
destination, 31
destruction, 19, 27, 105
detail, 9
detection, 19
deterrent, 88
detract, 29, 34
devaluation, 50, 52, 53, 57

development, 9, 20, 26, 28, 34, 38, 45–49, 53, 55, 64, 84, 87, 91–94, 97–99, 108, 109, 112, 114, 115, 118, 119, 124, 125, 127, 128, 130, 132, 136, 137
dialogue, 55, 69, 75, 76, 89, 90
diesel, 81
difference, 17, 102, 108, 116, 130
difficulty, 17
digestion, 123
dioxide, 8, 88, 107, 123, 125, 129
direction, 61, 79, 106, 116
dirt, 80
disadvantage, 103
discomfort, 70, 71
discourse, 13
discussion, 73
dismantling, 82, 84, 86
disorient, 20, 24
disorientation, 12
dispelling, 67, 92
displacement, 16, 23, 105, 106
displeasure, 71
disposal, 13, 86, 136
disruption, 19, 20, 23, 25, 27, 32, 33, 68, 99
distance, 38, 53, 68, 69, 136
distinction, 102
distortion, 50
distress, 69, 70
distribution, 136
district, 109, 124
disturbance, 12, 16, 17, 22, 23, 27, 32, 68, 98
diversification, 47
diversity, 32
diving, 24
downtime, 80, 81, 112

drag, 112, 116
dredging, 24, 25
drilling, 108, 109
drive, 44, 78, 113, 115, 123
drivetrain, 78
driving, 24, 31, 44, 47, 48, 100, 128, 137
drone, 67
durability, 78
dynamic, 113

Earth, 24, 107, 108, 119, 125, 129
earth, 103
echolocation, 17
ecology, 28
economy, 13, 29, 46, 50, 75, 84, 86, 130
ecosystem, 15, 16, 19
education, 46, 53, 57, 75
effect, 16, 26, 46, 53, 98, 119
effectiveness, 8, 28, 48, 102, 113, 115
efficiency, 4, 13, 43, 44, 76, 80, 84, 86, 108, 111–113, 118, 120, 122, 124, 129, 131, 134–136
effort, 131, 137
electricity, 1, 9, 10, 12, 13, 42–45, 47, 59–62, 64, 66, 81, 87, 88, 91, 93, 97, 99–107, 109, 114, 115, 120, 123–125, 129
electronic, 130
emission, 124
emphasis, 93
employment, 45, 46, 81
encroachment, 29, 53
end, 13, 24, 82
energy, 1–3, 5, 6, 8–14, 17–20, 22, 23, 26, 28, 30–35, 37–39,

42–51, 53–55, 57, 59–64, 66, 67, 70, 73, 75–79, 81, 82, 86–89, 91–94, 96–109, 111–115, 119, 120, 122–137
engagement, 14, 23, 34, 57, 67, 70, 73, 86, 88, 90, 92, 94
engineering, 9, 100
England, 34
enjoyment, 68
environment, 2, 19, 20, 32, 53, 69, 72, 83, 93, 97, 104, 129
equipment, 37, 38, 41, 45, 46, 81, 112, 129, 130
erosion, 82
establishment, 46, 124
Europe, 94, 99, 100
evidence, 53, 93
example, 24, 25, 29, 31, 32, 43, 44, 61, 63, 84, 93, 99, 100, 103, 111
exception, 89
execution, 38
expansion, 48, 49, 91, 115, 136
expenditure, 18
expense, 34
experience, 15, 29, 31, 32, 40, 50, 70, 72, 73, 78, 88, 96, 136
expertise, 40, 45, 66, 94, 130
exploration, 76, 112
exposure, 21, 68, 71, 98
extent, 39, 40, 53
extraction, 8, 107, 114

face, 9, 10, 17, 73, 89, 99, 105, 109, 131, 137
factor, 50, 93, 103, 105, 107, 115
failure, 77, 78, 92
fairness, 47

Falmouth, 52
family, 92, 103
fare, 136
farm, 26, 34, 38, 46, 53, 55, 69, 70, 75, 92, 94, 99, 100
fatigue, 68, 77
feasibility, 39, 66, 99, 115
feature, 5
feed, 48, 91
feeding, 16, 24, 25, 27
feedstock, 124, 125
feeling, 69, 71
fiber, 111
field, 1, 101, 104, 118
finance, 46
Finland, 124
fish, 12, 16, 20, 24, 25, 100, 105
fishing, 25, 32, 94
fit, 43
flexibility, 60, 102, 107–109
flicker, 26, 52
flight, 15
flow, 43, 65, 105, 106, 108, 119, 125
flowing, 105
fluctuate, 10, 61
flying, 12, 15, 19
focus, 27, 48, 84, 130, 134, 136
following, 20, 39, 41, 71, 76, 89, 104, 131
food, 16, 19, 25, 123
footprint, 88, 99, 103, 123, 125
foraging, 16, 17, 19, 23
forecasting, 12, 59, 61, 66
forefront, 99
forestry, 123, 124
form, 97, 113, 123
fossil, 1, 8–10, 12, 19, 20, 44, 45, 49, 87, 88, 99, 107, 108, 112, 114, 123–125, 129, 135

Index 145

foster, 55, 67, 69, 74, 90
foundation, 37, 38, 46, 66, 76
fragmentation, 17, 27, 98
framework, 66
frequency, 13, 39, 42, 44, 60, 68, 70–73, 88, 93, 106, 120
friendliness, 8
frustration, 68
fuel, 9, 12, 19, 44, 49, 87, 88, 107–109, 112, 114
functionality, 40, 132
future, 6, 10, 14, 19, 23, 28, 31, 33, 35, 39, 45, 47, 49, 53, 61, 64, 67, 73, 75, 76, 86, 87, 89, 92, 94, 100, 104, 107, 109, 113, 115, 118, 119, 122, 123, 125, 127, 128, 130, 131, 134–137

gain, 92
gap, 49, 75, 97
gas, 8, 26, 49, 87, 97, 108, 114, 122, 123, 125, 129, 130, 135, 136
gasification, 124
gear, 80
gearbox, 1, 78
generation, 9, 10, 13, 20, 22, 23, 26, 28, 32–34, 42–44, 49, 50, 55, 57, 60–62, 64, 66, 70, 73, 87, 88, 93, 100, 102–104, 106–109, 112, 113, 115, 123–125, 129
generator, 1, 78, 79, 87, 111, 113
geothermal, 8, 107–109
giant, 29, 68
Giromill, 116
government, 34, 47–49, 63, 65
grain, 87, 105

greenhouse, 8, 9, 19, 26, 49, 87, 97, 102, 103, 107, 108, 122, 123, 125, 129, 135, 136
grid, 10, 12, 14, 24, 38, 39, 42–45, 59–61, 64, 66, 67, 81, 91, 92, 96, 97, 104, 106, 107, 115, 120, 122, 136
growth, 9, 44–49, 75, 94, 99, 115, 122, 123, 130
guide, 22, 25, 26

habitat, 16, 17, 19, 20, 23, 24, 26, 27, 32, 98, 99, 105
hand, 19, 28, 38, 50, 102, 105–108, 113, 126
harm, 20, 26, 28, 53, 88
harmony, 33
harness, 1, 79, 92, 102, 111, 113, 115, 116, 125
harvesting, 124
hazard, 19
head, 82
health, 9, 11, 13, 21, 23, 25, 50, 52, 53, 68–73, 75, 83, 88, 93, 135, 136
healthcare, 46
hearing, 24
heart, 71, 111
heat, 107, 108, 123, 125, 129, 134
heating, 61, 109, 123, 124
height, 29, 30, 99
help, 2, 10, 17, 23, 25–28, 30, 32, 38, 47, 48, 53, 54, 61, 67, 69, 77–79, 81, 88, 91–93, 97, 98, 100, 107
heritage, 29–31, 33, 34
hiking, 29, 32
hillside, 29
history, 33, 35

home, 103
hope, 137
horizon, 15, 29, 31, 112
hospitality, 46
hour, 67
housing, 1
hub, 1, 55
hum, 19, 32, 68, 70
human, 9, 11, 21, 23, 25, 29, 35, 70, 71, 83
humanity, 33
hunting, 24, 26
hybrid, 78
hydro, 8, 10, 100
hydrogen, 107
hypertension, 71

ice, 80, 129
identity, 30, 33
impact, 10–15, 18, 20, 23, 25–34, 38–40, 50, 52, 53, 55, 56, 60, 65, 67–71, 73, 75, 76, 84, 88, 91–94, 96–100, 105, 107–109, 111, 112, 114, 129, 133, 135, 136
implementation, 57, 64, 66, 70, 73, 75, 81, 112
importance, 8, 10, 11, 28, 32, 34, 47, 49, 52, 61, 70, 84, 92, 94, 129, 131, 135, 137
imposition, 30
improvement, 136
increase, 38, 61, 71, 94, 129
independence, 130
individual, 9, 27, 70, 93, 103, 108, 109, 136
industry, 44–46, 48, 49, 69, 84, 86, 89, 97, 99, 113
inefficiency, 129

influence, 4, 50, 93
influx, 46
information, 69, 74, 75, 90
infrastructure, 9, 19, 30, 32, 37–39, 45–47, 59–61, 66, 86, 89, 91–93, 102, 106, 109, 114, 120, 124, 135–137
injury, 15, 19, 97
innovation, 26, 28, 44, 45, 85, 112, 115, 130
input, 129
insect, 27
insomnia, 68
inspection, 46, 79, 80
instability, 59
installation, 16, 17, 24, 26, 27, 29, 37–39, 43–45, 52, 56, 65–67, 104, 108, 111
instance, 24, 25, 30, 32, 80
insulation, 134
insurance, 38
integration, 14, 32, 42, 44, 45, 59–61, 64, 91, 97, 106, 107, 112, 115, 120, 122, 124, 136, 137
integrity, 29, 30, 34, 112
interaction, 21
interest, 48, 71, 87, 96, 130
intermittency, 10, 12, 13, 42–44, 59, 62–64, 97, 100, 101, 103, 105
interplay, 55
introduction, 29, 33, 70
intrusion, 33
intrusiveness, 31
investment, 37, 40, 47–49, 91, 93, 103, 106, 107, 115, 133
involvement, 23, 31, 81
ion, 43, 44, 63

irrigation, 134
irritability, 68, 71
issue, 10, 20, 22, 26–28, 30, 42, 53, 62, 93, 100, 135

job, 9, 44–47, 49, 50, 75, 99, 130
journey, 76
jurisdiction, 38

Kennedy, 92
knowledge, 28, 94, 104, 130

label, 129
lack, 86, 130
lake, 31
land, 17, 19, 27, 37, 38, 65–67, 80, 90, 91, 98, 102, 103, 114, 124, 125
landfill, 123
Landowner, 55
landowner, 53, 55
landscape, 5, 28, 29, 53, 64, 91, 92, 98, 111, 119, 134
landscaping, 134
layer, 20, 114
laying, 24
layout, 45, 75, 132
lead, 16, 17, 21, 23, 50, 56, 59, 68, 71, 76, 77, 84, 85, 90, 100, 105, 111, 130
lease, 54, 55, 90
led, 8, 61, 69, 111, 129, 132
leeward, 116
leg, 114
legislation, 55
length, 111
lidar, 26
life, 12, 13, 20, 24–27, 43, 46, 52, 54, 68–71, 82, 98, 100
lifecycle, 86

lifespan, 13, 37, 44, 102–104, 106, 107, 109
lift, 112, 116
light, 16, 130
lighting, 25, 124
likelihood, 26
limit, 68
link, 13, 71
lithium, 44, 63
litigation, 91
location, 34, 38, 39, 50, 60, 70, 108, 109
longevity, 13, 30, 39, 43, 78, 79, 82, 103
look, 2, 47, 76
loss, 27, 61, 98, 134

machinery, 46
maintaining, 10, 30, 32, 53, 60, 82
maintenance, 9, 13, 38–46, 77–82, 102, 103, 106, 111, 112
making, 16, 19, 30, 35, 37, 39, 43, 47, 48, 55, 61, 75, 88, 90, 100–102, 106, 108, 122, 123, 125, 135
mammal, 12
man, 29
management, 12, 13, 18, 28, 38, 42, 46, 59–61, 82–86, 98, 106, 120, 124, 125, 133, 135
manner, 28, 82, 86
manufacturer, 39
manufacturing, 9, 44–46, 103, 108, 111
manure, 123, 124
market, 50, 91
Massachusetts, 31, 52, 92
material, 77, 84, 119
mating, 24

matter, 123, 124, 129
mature, 48, 115, 128
means, 43, 82, 102
meeting, 87, 107, 109
metal, 86
method, 104
migration, 18, 19, 22, 24–26, 97
million, 99
mind, 29, 49
mindedness, 76
mindset, 84
mining, 46, 103
mitigation, 10, 17, 18, 22, 26–28, 33, 35, 53, 54, 56, 57, 65, 69, 71, 73, 74, 88, 93, 94, 100
mix, 61, 119, 120, 124, 125
mode, 136
modeling, 60
modernization, 91
module, 119
moisture, 124
momentum, 99
money, 46, 130, 131
monitoring, 17, 23, 25, 27, 28, 40, 42, 43, 46, 69, 70, 77, 79–82, 112, 124
monument, 34
moon, 125
mooring, 115
mortality, 12, 22, 23
motion, 1, 77, 114, 116, 126
moving, 19
myriad, 129

nacelle, 1
nature, 10, 12, 14, 29, 32, 40, 42, 48, 61, 66, 68, 97, 103, 106
navigation, 12, 17, 20, 24, 67

necessity, 8
need, 8, 12, 18, 23, 26, 30, 34, 42, 43, 48, 51, 53, 59, 60, 63, 66, 76, 78, 80, 82, 84, 87, 89, 91, 93, 100, 102–104, 106, 109, 111, 112, 114, 121, 124, 129, 134, 135, 137
negotiation, 55, 65, 91
neighboring, 53
net, 123
network, 45, 136
night, 16, 26
nighttime, 71, 103, 104
nitrogen, 124, 129
noise, 12, 13, 16, 17, 19–25, 27, 31, 32, 38, 52, 53, 56, 67–73, 75, 88, 90, 93, 98–100, 111, 114
number, 1, 9, 12, 25, 29, 31, 45, 53, 80, 84, 136

obstacle, 15, 100
ocean, 113, 125, 126
oil, 8, 87, 114, 125, 130
one, 9, 28, 37, 44, 47, 61, 65, 78, 80, 88, 98, 99, 113
onshore, 99, 100
operating, 12, 37, 39, 60, 80, 120
operation, 4, 12, 13, 17, 19, 20, 24, 26, 32, 39, 43–46, 60, 61, 65, 77, 79, 92, 98, 100, 108, 112, 114, 116
operator, 66
opinion, 13, 66, 93
opportunity, 100, 137
opposition, 10, 13, 31, 34, 55, 73, 75, 87–90, 92, 109
optimization, 23

Index 149

option, 45, 87, 104, 107, 136
order, 56, 65, 89
orientation, 24, 79
origin, 87
other, 1, 16, 19, 25, 32, 38, 46, 50, 60, 61, 64, 66, 90, 93, 96, 102–108, 113, 115, 120, 126, 130, 136
outlook, 128
output, 12, 42, 59, 61, 66, 106, 108, 109, 113, 129
outreach, 130
overview, 94
ownership, 65, 75, 81, 136

pad, 38
panel, 103
part, 1, 14, 15, 30, 61, 111, 125
participation, 55, 92
particulate, 124, 129
party, 40
passage, 24
passenger, 136
past, 33
path, 19
pathway, 10
peak, 19, 26, 27, 106, 112
pedestrian, 136
penetration, 45
people, 13, 61, 98, 130, 136
percentage, 120
perception, 14, 24, 53, 66, 70, 73, 75, 76, 93, 94
performance, 4, 13, 39, 40, 42, 43, 46, 65, 71, 76–81, 112, 114, 115, 129, 132, 134
period, 40
person, 98
personnel, 41, 43

perspective, 89
phase, 46, 66
phenomenon, 50
photosynthesis, 123
photovoltaic, 119–122
pile, 24, 100
pitch, 27
place, 37, 40, 56, 71, 91, 104
placement, 19, 20, 22, 30, 65, 67, 98
plan, 40, 81, 82, 92
planet, 129
planning, 14, 16, 20, 27, 30–33, 48, 52, 60, 61, 66, 67, 80, 84, 86, 88, 90, 92, 94, 124
plant, 106, 123, 125
plate, 108
platform, 114, 115
player, 8
point, 31, 88
pollution, 12, 13, 20, 22–24, 27–32, 53, 56, 67, 68, 70, 71, 73, 88, 90, 93, 98, 99, 102, 103, 123, 124, 129, 135
pool, 81
popularity, 11, 97
population, 16, 25
portfolio, 91, 107
portion, 104
positioning, 113, 134
possibility, 30
potential, 6, 9, 10, 13, 15, 16, 18–21, 23, 25–28, 30–33, 38–40, 42, 44, 49, 50, 53, 55, 59, 64–66, 70–73, 75, 79, 81, 82, 84, 88–91, 93, 96, 97, 99, 101, 106, 113–115, 122, 123, 125, 128, 130, 137

power, 1, 8–10, 12–14, 18, 19, 24, 31, 34, 42–44, 47–50, 55, 59–61, 66, 79, 81, 87, 88, 91, 99, 100, 102–109, 111–113, 120, 123–125, 129, 136
practice, 104
predictability, 47
premium, 47
preparation, 37, 39
presence, 16, 17, 25, 26, 29, 31, 32, 50, 53, 65, 68, 81, 88, 98
preservation, 17, 20, 26, 29–31, 33–35, 55
pressure, 17, 19, 27, 71, 116
prey, 17, 20, 24
price, 23, 43, 93, 130
principle, 1, 3, 113, 123
priority, 20, 25
problem, 19, 20, 40, 67
process, 3, 30, 38, 55, 65, 66, 75, 82–84, 86, 90–92, 94, 108, 112, 123, 124
production, 17, 46, 48, 81, 84, 97, 103, 106, 108, 111, 112, 125, 136
productivity, 68
professional, 47
profit, 47
profitability, 91
program, 87
progress, 98
project, 38, 43, 46, 47, 63, 64, 66, 73, 75, 84, 89–94, 99, 108, 109, 112, 115
proliferation, 13, 15
promise, 100, 114, 125, 127
property, 46, 50–57, 67, 88, 90, 92
protection, 18, 28, 31, 38, 57, 78, 79, 130
provision, 70
proximity, 13, 27, 50, 53, 66, 68–70, 88
public, 10, 13, 14, 46, 48, 50, 55, 69, 73–76, 92–94, 130, 136, 137
pull, 125
purchase, 39
purchasing, 37, 137
pursuit, 28
pyrolysis, 124

quality, 9, 27, 46, 52, 54, 68–71, 98, 120, 131
quest, 131
quietude, 32

radar, 26, 66
range, 41, 45, 68, 70, 103, 106, 124
rate, 71
ratio, 111
reading, 104
recourse, 55
recreation, 29, 31–33, 68, 98
recycling, 13, 81, 84, 86, 103, 136
reduction, 75, 130
refuge, 29
region, 27, 31, 44, 63
regulation, 44, 69, 70
relationship, 72, 90
release, 8, 43, 97, 103, 107, 124
reliability, 12, 42, 44, 45, 60, 61, 66, 78, 82, 103, 108, 115
reliance, 10, 43–45, 99, 112, 124, 125, 129, 135
reminder, 61
removal, 82, 86

Index

renewal, 48
repair, 40, 46, 77, 79, 112
replacement, 40, 77, 80
reporting, 52
research, 13, 17, 20, 23, 28, 45, 49, 71, 73, 85, 87, 89, 115, 118, 119, 124, 125, 128, 136
reserve, 44
resilience, 61, 79, 130
resistance, 13, 80, 93, 137
resolution, 81
resource, 9, 59, 65, 81, 84, 86, 89, 108, 109, 119, 133, 135
response, 44, 60, 64, 75, 114, 115
responsibility, 81, 98
restoration, 27, 86
result, 15, 16, 18, 26, 27, 29, 32, 42, 47, 50, 52, 103, 104, 114
return, 80, 103
revenue, 29, 46–48, 91, 93
Rhode Island, 100
rise, 15, 125, 129
risk, 15, 16, 20, 22, 24–26, 42, 44, 47, 49, 79, 91, 93, 97
river, 105, 106
road, 136
role, 6, 8, 13, 17, 25, 28, 32, 45, 47–50, 53, 64, 66, 67, 73, 80–82, 87, 97, 101, 115, 125, 128, 131, 132, 134, 135, 137
rooftop, 103
roosting, 17, 27
rotation, 68, 113
rotor, 1, 76–79, 87, 111–113
row, 29
run, 130
rush, 67

safety, 41, 67, 76, 79, 80, 92, 114
salt, 80
satisfaction, 32
saving, 130, 131, 133
scalability, 8, 9, 43, 103, 104, 108, 111
scale, 2, 24, 43, 44, 55, 59, 60, 63, 87, 93, 100, 103, 105, 108, 109, 120
scarcity, 81, 114
scheduling, 81
Scotland, 115
screening, 53
sea, 24, 92, 99–101, 114, 129
seabed, 24, 98, 113, 115
seagrass, 25
section, 8, 15, 20, 24, 26, 28, 31, 39, 42, 45, 50, 67, 70, 73, 76, 80, 89, 94, 97, 99, 104, 107, 111, 113, 115, 123, 125, 131, 135
sector, 9, 45–47, 101
security, 88, 122, 130
selection, 32, 33, 73
self, 2
semi, 114
semiconductor, 119, 120
sense, 30, 33, 68, 69, 81
sensitivity, 70
sensor, 112
sentiment, 92
serenity, 32
service, 40
servicing, 80
set, 13, 19, 34, 80, 97, 101, 116
setback, 27, 52, 53, 70, 93
setting, 37, 69, 103
shade, 134
shaft, 78

shape, 19, 112
share, 60
sharing, 47, 93, 94
shelter, 19
shift, 13, 75, 130, 137
shipping, 25, 112
shopping, 50
shortfall, 12
sight, 28
significance, 29, 34, 93
silicon, 119
simulation, 60, 112
site, 32–34, 37–39, 45, 66, 73, 80, 86
siting, 19, 20, 22, 25, 26, 65, 67, 98, 109
size, 12, 16, 21, 26, 29, 31, 43, 53, 108, 111–113
skepticism, 73
skyline, 29
sleep, 23, 68–71, 98
society, 33, 55, 135, 137
sodium, 43
software, 112
soil, 66, 124
solace, 32
solar, 8, 81, 87, 102–104, 119–122, 130, 136
solution, 6, 10, 19, 26, 39, 42, 81, 113, 119, 122, 124, 125
sonar, 17
sound, 12, 20, 24, 68, 70, 88
soundscape, 20, 25, 72, 73
source, 1, 8–11, 23, 31, 39, 46, 59, 61, 68, 70, 73, 88, 94, 97, 99, 101, 102, 105–107, 109, 123, 125
South Australia, 44, 63
space, 65, 102–104

spar, 114, 115
spawning, 25
species, 16, 17, 19, 20, 22, 24–28, 32, 65, 88, 97
speed, 12, 21, 27, 42, 59, 61, 68, 78, 97, 106, 108
spending, 47
spin, 1, 107
spinning, 15, 17–19, 24, 26, 27, 32, 44
spot, 65
stability, 12, 38, 43–47, 60, 61, 64, 66, 77, 97, 106, 114, 115
staff, 41, 42, 46
stakeholder, 31, 73, 88, 92, 94
standardization, 86
start, 75, 92
state, 55, 61
steam, 107, 123
steel, 1, 78, 84
step, 3, 38, 53, 73, 75, 113, 135
stewardship, 16, 93
stillness, 32
stimulation, 130
stimulus, 45
storage, 10, 13, 42–45, 59, 60, 62–64, 66, 97, 99, 100, 103, 104, 106, 112, 122, 125
store, 43, 97, 104
strain, 12, 61, 129
strategy, 125
stream, 47, 48
strength, 111
stress, 19, 23, 24, 68–71, 98
strike, 18, 22, 26, 28, 30, 32, 33, 35, 55, 67, 70, 89, 99
structure, 1, 68, 77, 113, 114
study, 34, 52, 69, 84, 94

Index 153

subject, 13, 59, 105, 106
success, 13, 27, 44, 55, 67, 79, 82, 94
suction, 115
suitability, 66
sulfide, 107
sulfur, 43, 129
summary, 109
summer, 12, 134
sun, 104, 119, 125
sunlight, 102, 119, 120, 123
supply, 10, 12, 13, 42–45, 59, 61, 62, 64, 81, 97, 100, 103, 105, 107, 112, 124
support, 13, 46–49, 64, 66, 69, 70, 75, 77, 92, 94, 100, 114, 130
surface, 103, 107, 126
surge, 45
surrounding, 15, 16, 18, 19, 24, 32, 47–49, 54, 93, 98, 134
survival, 18–20, 25, 27
sustainability, 11, 39, 45, 48, 76, 98, 133
Sweden, 124
symbol, 98
syndrome, 67
system, 12, 43, 60, 61, 78, 79, 81, 103, 104, 106, 119, 120

tag, 130
tax, 46–48, 130, 137
technology, 14, 16, 30, 32, 39, 40, 43–45, 49, 86, 87, 96, 100–102, 111–113, 115, 122, 129, 132, 136, 137
temperature, 81, 105, 120, 129
tendency, 27
tension, 114

term, 13, 25, 28, 37, 39, 40, 42, 46–49, 71, 91, 98, 133
terrain, 37, 66
testament, 44
Texas, 55, 61
the Block Island Wind Farm, 100
the Thames Estuary, 99
the United Kingdom, 99
The United States, 87
the United States, 48, 94, 100
thing, 65
threat, 12, 24, 32, 33, 83, 88, 97
ticketing, 136
time, 10, 13, 40, 48, 60, 61, 68, 75, 80, 81, 90, 103, 112, 119, 129, 131, 136
tip, 134
today, 129, 131
tomorrow, 135
tool, 9
topic, 8, 20, 70, 73, 76, 88, 94
topography, 106
tourism, 12, 28–32, 46, 68, 69, 88, 94, 98
tourist, 30–33
tower, 1, 76–78
town, 52
tracking, 136
traffic, 67, 136
training, 41, 42, 80, 81
tranquility, 31–33
transition, 6, 8, 20, 26, 28, 34, 44, 46, 86, 87, 96, 101, 107, 115, 122, 123, 128, 130, 137
transmission, 19, 27, 38, 60, 61, 66, 91, 93
transparency, 91
transport, 66, 80, 135

transportation, 38, 45, 50, 80, 86, 108, 112, 123–125, 135–137
travel, 80, 135
treatment, 50
trip, 80
troubleshooting, 46
trust, 13, 90, 92
truth, 19
turbine, 1, 4, 13, 15, 16, 19–21, 23, 25–27, 32, 39, 40, 42, 45, 46, 65–71, 73, 75, 77–82, 84, 87–91, 93, 97, 105, 111–115, 123
turbulence, 65, 106, 112, 114
turn, 29, 49, 113
type, 43, 102

U.S., 100
UK, 34
uncertainty, 48, 91
understanding, 19, 26, 55, 67, 89, 92
unease, 68
unit, 1, 43, 44, 111
unpredictability, 12, 48
unpreparedness, 61
upgrade, 60
upright, 114, 115
USA, 31
usage, 81, 104, 130, 134
use, 9, 27, 38, 40, 60, 65–67, 81, 82, 103, 104, 106, 112, 113, 129–131, 136, 137
utility, 87, 108, 131
utilization, 44, 60, 82, 99, 102, 112, 136

value, 29, 33, 34, 51, 53, 91, 103

variability, 12, 13, 59–61, 66, 106, 108, 124
variety, 120
veneer, 11
viability, 39, 42, 44, 48, 64, 91–94, 99
vibration, 19, 81
victim, 24
view, 29
village, 29
visitor, 29, 32
vitality, 68
voltage, 42, 60, 120

walkability, 137
walking, 136
warming, 129
warranty, 38, 40
waste, 13, 50, 82–86, 105, 123–125, 129, 131, 135
wastewater, 125
water, 81, 87, 105–107, 113, 123–125, 134
waterfowl, 16
wave, 125–128
way, 76, 99, 104, 109
weather, 12, 61, 80, 102, 103, 129
web, 18, 25
weight, 111
well, 21–23, 41, 49, 57, 68–71, 73, 93, 94, 109
whole, 33
wildlife, 10–12, 14, 19, 20, 22, 23, 25, 26, 28, 32, 33, 38, 65, 75, 90, 93, 97–100, 105
wind, 1–4, 8–34, 37–40, 42–57, 59–94, 96–109, 111–116, 119, 126, 130, 136
winter, 61, 134

wood, 124
work, 3, 46, 67, 75, 89, 91, 116, 135, 137
workforce, 81
world, 5, 9, 10, 28, 31, 32, 44, 47, 69, 89, 99, 112, 127, 129, 137

Xeriscaping, 134

Yaw, 79
year, 15, 99

zoning, 30, 66, 90

Milton Keynes UK
Ingram Content Group UK Ltd.
UKHW020840300824
447605UK00010B/256